THE POLITICAL ECOLOGY OF BANANAS

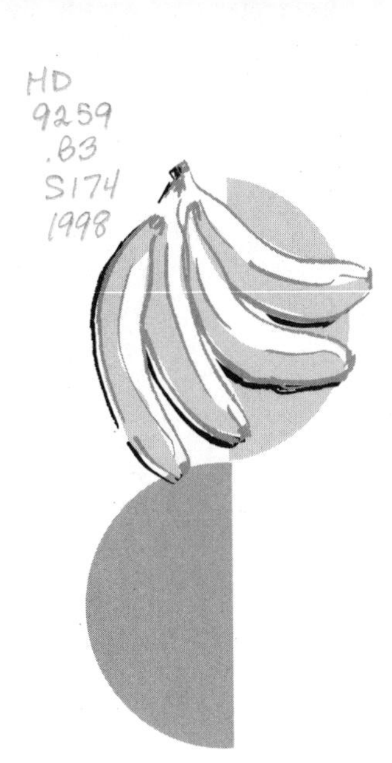

Lawrence S. Grossman

The POLITICAL ECOLOGY of BANANAS

Contract Farming, Peasants, and Agrarian Change in the Eastern Caribbean

The University of North Carolina Press
Chapel Hill and London

Manufactured in the United States of America

The paper in this book meets the guidelines for permanence
and durability of the Committee on Production Guidelines
for Book Longevity of the Council on Library Resources.

Library of Congress Cataloging-in-Publication Data

Grossman, Lawrence S., 1948–
The political ecology of bananas: contract farming,
peasants, and agrarian change in the Eastern Caribbean /
Lawrence S. Grossman.
p. cm.
Includes bibliographical references and index.
ISBN 0-8078-2410-0 (cloth: alk. paper).
ISBN 0-8078-4718-6 (pbk.: alk. paper)
1. Banana trade—Saint Vincent and the Grenadines. 2. Contracts,
Agricultural—Saint Vincent and the Grenadines. I. Title.
HD9259.B3S174 1998
338.1′74772′09729844—dc21 97-40002
CIP

02 01 00 99 98 5 4 3 2 1

To Mina

Contents

Maps and Figures

Maps

Figures

Tables

Preface

My understanding of the significance of changes in the banana industry in the Windward Islands of the Eastern Caribbean—and of agrarian transformations more generally—stems not only from long-term field research in developing countries but also from my brief experience as an industrial laborer at a paper factory in Philadelphia in the summer of 1967. My assignment was simple. For eight hours I had to stand in front of a large machine that held a very thick roll of heavy construction paper approximately nine feet across. As the machine continuously dispensed the paper from the roll, two razor-sharp blades cut the material into three equal-sized sheets, which came flying onto a metal table in front of the machine. As the speed of the machine was quite rapid, the sheets would end up in a haphazard heap unless someone was there to stack them into a neat pile. That is where I came in. Thus, at the table I had to position each sheet quickly and carefully into a neat pile that could then be bundled for shipment by someone else in the factory. Being new, I could handle only one pile at a time. The person working next to me at the table, a veteran of eight years at the plant, was able to stack two piles at a time, one with each hand. That is what we did for eight hours. Although factory employees have jobs that certainly vary in their content and scope, what I did was typical for what the industrial literature describes as a "deskilled" factory worker in a mass production setting—performing only a very small portion of the overall production process in a monotonous, repetitive task that did not require—and indeed my employers did not want—creativity. Others in the company had designed the industrial process; I simply carried out their directives. My experiences at a chemical factory later that summer provided relatively similar, though somewhat more hazardous, experiences.

What does this have to do with agriculture? A major theme today in the agrarian literature generally and on contract farming in particular is to examine changes in contemporary agriculture through the lens of the industrial restructuring literature because there are supposedly significant parallels between the two realms. This book argues that such an endeavor is misguided. Rather, I assert that labor processes in agriculture and industry are fundamentally different. Agriculture requires a much more creative set of responses in the production process than does industry in the setting of mass production and deskilled workers; it is certainly not a repetitive, monotonous activity. Specifically, the environmental context of peasant agriculture—which creates conditions inherently variable in space and time—makes homogenization of the production process in such endeavors as contract farming much more problematic and difficult than in industrial settings. Exploring the significance of such differences is a major theme of this book.

My first major experience in fieldwork was in the highlands of Papua New Guinea in the late 1970s. Based on that research, I became convinced that commodity production entailed significant negative impacts for small-scale farmers, especially in relation to food production. I became interested in contract farming because it represents an intensified and more regulated form of commodity production. Thus, I devised a project to study the impact of contract farming of tobacco in western Kenya in the mid-1980s. I conducted a brief, one-month survey of contract-farming systems in that country in 1986 to help gain some perspective on the range of systems practiced there. But political turmoil at the time prevented me from returning for longer-term research.

At the suggestion of Bon Richardson, a colleague in my department, I then shifted my focus to St. Vincent in the Eastern Caribbean, where I ended up studying the banana industry and its system of contract farming. It turned out to be a fascinating experience. One comment from a farmer a few days after I arrived made me realize quite early that my research would force me to reconsider some of my previously held convictions. I mentioned to the farmer that I had read that one of the major problems in the English-speaking Caribbean was inadequate food production, and that in the case of St. Vincent and the other banana-growing Windward Islands, banana production was mainly responsible for the problem. He looked at me with a bit of puzzlement and then proclaimed that the real problem in St. Vincent was that farmers produced too much food! I knew that I would have much to learn.

After spending two weeks in Barbados examining secondary materials, I arrived in St. Vincent in August 1988 and stayed there until August 1989. Most of my research focused on one banana-growing village community, though I traveled widely on the island and visited St. Lucia, another of the Windward Islands, several times to obtain comparative material. I returned to St. Vincent and the village for brief visits in 1990, 1991, 1992, 1994, and 1995, which helped me gain a longer-term perspective on changes in the industry. This work was supplemented by a visit to the United Kingdom in the summer of 1990, when I was able to examine historical records at various institutions and to meet with officials at Geest, at the time the exclusive marketer of Windward Islands bananas. That same year I was also able to explore the records on British foreign aid to the Windwards banana industry at the British Development Division in the Caribbean on Barbados.

One cannot study the banana industry and fail to comment on the current threats that it faces. Recently, the World Trade Organization (WTO) ruled in favor of the United States on its complaint that the market protection provided by the European Union (EU) to banana exporters such as St. Vincent and the other Windward Islands violates international trade rules. The WTO decision may result in dismantlement of the protection on which the Windward Islands depend. It is unlikely that they will be able to survive in an unregulated market against much larger and more powerful, predatory transnationals—Chiquita, Del Monte, and Dole—which produce and market bananas from Latin America. Such an outcome will certainly be a tragedy for the banana growers who depend so much on the export of the crop for their survival. Banana production, while being very demanding of labor and resources, has nonetheless been a major reason for substantive improvements in living standards on these islands over that last forty years. The Jamaican prime minister Percival Patterson's comments accurately reflect the importance of the banana industry to the Windwards and other Caribbean producers: "Bananas are to us what cars are to Detroit" (James 1997: 4). It is inexcusable and morally wrong for the United States to use its considerable might to destroy the livelihoods of so many small-scale farmers who have few alternatives for earning a living, especially when the United States itself exports no bananas. To put the glaring inequalities of the situation into perspective, St. Vincent earned US$19.4 million in 1996 by exporting bananas—the lifeblood of its economy—whereas in the following year the United States spent over US$40 million just on President Clinton's inauguration party.

Acknowledgments

I was very fortunate to have selected the village of Restin Hill as a research site. Whether walking up steep hills to examine gardens, discussing the fine points of banana cultivation, enjoying a beer with friends in one of the rum shops, or sharing a typically delicious meal at someone's home, the people of the community always made me feel at home and were most supportive of my research. They provided a year's worth of research materials and a lifetime of memories. For this I am most grateful. There are numerous individuals whom I would like to thank specifically. My good friend and tireless field assistant, Chesley, was truly a great help in my research. Merlene and Wesmore and their children always made me feel as if I were a part of their family; I thoroughly enjoyed the many hours I spent in their gardens and their home. I shared similarly wonderful experiences with Jonathan and Thelma, Inez and Garvin, Miss Millie, Jean, Doris and Cornelius, and Ray and Homalita and their family. And Patrick, Marjorie, and Catherine were particularly effective teachers in the art of agriculture. I would also like to thank the other members of the sample for their help, especially Elsa, Cilter, Elgin and Elfrita, Elizabeth, Iona, Roy and Christina, Brother Ross and Janice, Irmilin, Estelle, and Francis and Lorna. Mrs. Robertson, Nappy, and Dexter were also very supportive and good friends.

Members of the St. Vincent Banana Growers' Association provided much help in my research. I have to single out Sylvester Vanloo, who from the beginning was always willing to teach me about the intricacies of banana production; fortunately for this research, he never tired of my endless questions. I also thank Mrs. Bess, Hugh Stuart, Henry Keizer, Vibert Williams, and Ashley Cain. And Mr. William Jack, a pioneer in the industry, kindly retold many stories about the history of producing bananas on the island. I also learned much from discussions about problems in the banana industry from officials at WINBAN (now WIBDECO) in nearby St. Lucia—Dr. Errol Reid, Mike Augustin, and Theresa Alexander-Louis.

The Ministry of Agriculture, Industry, and Labour on St. Vincent was similarly supportive. Reuben Robertson was very helpful, and I have to thank him, in particular, for suggesting that I conduct my research in Restin Hill. I had fruitful discussions with Philmore Isaacs about the issues of pests and pesticides. And I appreciate greatly all the assistance that I received from Tony Caesar, an agricultural extension agent, who provided many insights into Vincentian agriculture.

My research is indebted as well to many others on the island. I specifically thank Clive Bishop, Adrian Fraser, Kenneth Bonadie, and Ivo Sinson. And Selwyn Allen at the Statistical Office was most supportive of my efforts to obtain various data in the archives. The Organization for Rural Development kindly provided access to its facilities. I have also gained from numerous discussions with anthropologist Hymie Rubenstein.

The Center for Resource Management and Environmental Studies and Dr. Euna Moore at the University of the West Indies on Barbados, provided affiliation. I greatly appreciate Jeremy Collymore's help in establishing the linkage and in facilitating the beginning of my research. Also on Barbados, John Ferguson at the British Development Division in the Caribbean helped arrange my research at the organization's office.

Many at Virginia Tech were instrumental in encouraging this project. I have enjoyed my many talks about the Caribbean with Bon Richardson, who came up with the idea of my conducting research on St. Vincent. Jane Price and Vanessa Scott did a wonderful job in typing the manuscript, and John Boyer artfully created the maps.

I also thank several colleagues for their most helpful suggestions after reading the manuscript—Janet Momsen, Peter Little, and Abdi Samatar. And Elaine Maisner at the University of North Carolina Press has been a constant source of support.

The National Science Foundation, Geography and Regional Science (SES-8706689), and the National Geographic Society (3623-87) provided very generous funding for this research. The College of Arts and Sciences at Virginia Tech has also facilitated this work through supplemental funding on several occasions.

Most important, I thank my wife, Mina, for enduring a year apart and for being so supportive and understanding.

Abbreviations

ACP	Former colonies of EU countries in Africa, the Caribbean, and the Pacific that are part of the Lomé Convention
BAT	British American Tobacco Company
BDD	British Development Division
EC$	Eastern Caribbean dollar (EC$2.70 = US$1.00)
EC	European Community
ECU	European Currency Unit
EEC	European Economic Community
EU	European Union
GATT	General Agreement on Tariffs and Trade
GDP	Gross domestic product
IBD	Internal Buying Depot
MAFF	Ministry of Agriculture, Fisheries, and Food (United Kingdom)
NAFTA	North American Free Trade Agreement
NICS	Newly industrializing countries
SEM	Single European Market
SVG	St. Vincent and the Grenadines
SVBGA	St. Vincent Banana Growers' Association
WIBDECO	Windward Islands Banana Development and Exporting Company
WINBAN	Windward Islands Banana Growers' Association
WTO	World Trade Organization

THE POLITICAL ECOLOGY OF BANANAS

Introduction

Political Ecology and Contract Farming

"Bananas are like babies because of all the care you have to give them," declared a veteran female banana grower on the Eastern Caribbean island of St. Vincent. Another villager, standing close by in an adjoining banana garden, nodded her head in agreement and added, "When you plant banana, you have to live in it." These reflections portray the truly demanding nature of banana export production, which has been the pillar of the island's economy since the mid-1950s. From planting to harvesting, the crop absorbs the attention of thousands of farmers not only in St. Vincent but also on the other Windward Islands of St. Lucia, Dominica, and Grenada (Map 1). What makes banana production so interesting to the outside observer—and so burdensome for Vincentian farmers—is that it is technologically much more complex than any other form of agriculture ever practiced on the island. Moreover, the pace of technological change in the industry has accelerated over time in response to ever increasing demands from British capital, the state, and, most recent, the European Union (EU) for improvements in farm productivity and fruit quality.

Intervention by capital and the state in the peasant production process leading to technical innovations and labor intensification is characteristic of systems involving contract farming, which is a fundamental element of the banana industry in the Windward Islands. Contract farming is one of the most significant and powerful means by which peasants have been integrated into national and international commodity markets and agro-industrial complexes. The nature and structure of contract-farming systems vary widely from place to place, but a fundamental element is the vertical concentration of producers in which capital and the state attempt to supervise and condition the production patterns of growers (Bernstein 1979; Little and Watts 1994). Many researchers view this institutional form as an integral element in the "new international division of labor in agriculture" and the "new internationalization of agriculture" (Sanderson 1985a,b, 1986a,b; McMichael and Myhre 1991; Raynolds et al. 1993; Watts 1994a). It first became prominent in advanced capitalist countries but has spread rapidly in so-called developing countries in the post–World War II era (Glover

MAP 1. St. Vincent and the Grenadines

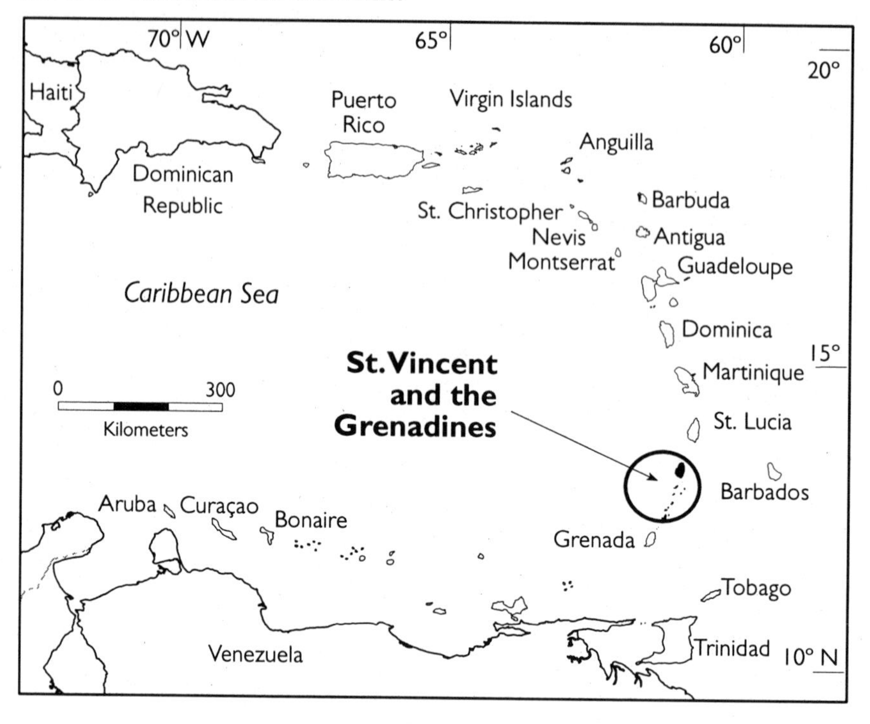

1984; Glover and Kusterer 1990; Little and Watts 1994). Thus, Vincentian banana growers share certain similarities with other contract farmers, including those who grow rice in The Gambia (Carney 1994), tea in Kenya (Buch-Hansen and Marcussen 1982), broccoli in Guatemala (Glover and Kusterer 1990), onions in Thailand (Laramee 1975), and grapes in Chile (Korovkin 1992).

In this book I employ the increasingly popular perspective of political ecology—which highlights the complex relations between political economy and patterns of resource use—to examine the institution of contract farming in the Windward Islands banana industry and its impacts on the peasantry in St. Vincent and the Grenadines.[1] My scope is broad, ranging from Vincentian households and their gardens to transnational corporate strategy, British policy, and the changing nature of the EU. Similarly, its time dimension is long, tracing the evolution of the political economy of land use on St. Vincent and the growth of the banana trade linking the Caribbean to the United Kingdom—both of which have significant implications for the contemporary functioning of contract farming in the Windwards banana industry. Although a few other studies have focused on banana production in the Windwards at various levels of analysis (for example, Marie 1979; Thomson 1987; Trouillot 1988; Nurse and Sandiford 1995; Welch 1996), none has considered the industry as incorporating contract farming, and hence they have failed to gain the analytical benefits from examining Windwards production in light of the rapidly growing literature on the topic.

The study of contract farming is particularly exciting because it provides a window on many themes of fundamental importance to the literature on economic development. In this book, I highlight three interrelated issues especially significant in the analysis of contract farming. The "labor question" concerns the extent and import of control exercised by capital and the state over the peasant labor process, an issue that is particularly relevant to contemporary debates about the applicability of the concept of "deskilling" to contract farming. The "food question" explores the impacts of contract farming for export on domestic food production and food import dependency, relationships that are much more complex than the literature indicates. Last, the "environmental question," reflecting the general trend of intensive agrochemical use in contracting, investigates the environmental consequences of banana production and the often misunderstood and inadequately explained problem of "pesticide misuse." This discussion of banana contract farming also has significance for understanding the contentious debates about the uniqueness of agriculture and the problems of applying generalizations developed to understand restructuring processes in industry to the analysis of patterns of agrarian change (see Goodman and Watts 1994; Jarosz 1996a). Furthermore, it is directly relevant to contemporary discussions about the globalization of agriculture. Before examining these issues in greater depth, I first provide an overview of the nature of contract farming, which will help place the Vincentian banana industry in comparative perspective.

Contract Farming

To understand why contract farming is becoming increasingly widespread in developing countries and to illustrate its unique characteristics, it must be distinguished from more traditional forms of commodity procurement—open-market sales and vertical integration (Minot 1986). In open-market sales, no preexisting agreements exist between buyers and sellers. Farmers have no guarantees concerning what prices they will receive or how much they can sell. Buyers, in turn, have to worry about the possible irregularity of supplies, a particular problem for crop processors whose factories must operate close to capacity to maintain profitability (Daddieh 1994); for exporters of off-season fruits and vegetables who have to ensure that supplies are available at the right time of the year (Raynolds 1994a,b); and for marketers of perishable fresh fruits and vegetables who need to supply their customers continually on a regular basis. Moreover, as buyers have no control over the production process in open markets, they have limited influence over crop quality, a concern in the context of an increasingly competitive and quality-conscious global marketplace. In contrast, in vertical integration, in which crop processors/marketers are also involved in direct production and other stages of the commodity chain, problems of regularity of supply and quality of output are lessened because these enterprises make their own decisions concerning scheduling and agronomic practices. But direct production, particularly for large-scale enterprises, also has drawbacks. Production risks associated with adverse weather and pests, problems of labor recruitment, motivation, and control, the threat of strikes and crop sabotage by disgruntled workers, and the possibility of their foreign-owned enterprises being expropriated all pose uncertainties for vertically integrated enterprises that can seriously undermine profitability.

Contract farming solves many of the problems inherent in both open-market sales and vertical integration. Although systems of contract farming exhibit a wide range of variation in their characteristics, two fundamental features are integral to all such enterprises—the provision of a guaranteed market and some control exercised by capital and the state over the production process (Watts et al. 1988). Thus, the purchaser—which can be a state agency, private local or foreign capital, or a public/private joint venture—agrees to buy the output of farmers before production begins, usually provides them with agricultural inputs on credit, advises and supervises them through its own agricultural extension service, and establishes mechanisms for price determination. In return for the guaranteed market, farmers are supposed to follow the procedures for cultivating and harvesting that are established in the contract and accept the advice provided by the firm's extension officers.

The system, in theory, has potential advantages for both sellers and buyers. Farmers benefit from a guaranteed market, which reduces the uncertainty associated with wide fluctuations in demand characteristic of open markets. Although methods of price determination vary considerably, price stability is also normally greater in contract farming. Moreover, small-scale farmers who previously could not afford to purchase yield-improving inputs, such as fertilizers, pesticides, and hybrid seeds, can now obtain them on credit through such schemes.

Substantial advantages also accrue to buyers. The riskiest and least profitable stage of the commodity chain is left to farmers, and the threat of expropriation is negated (Goldsmith 1985). Contracting provides a more regular and dependable supply for processing, marketing, and exporting than open-market purchases. Contract farmers are more highly motivated to produce high-quality output than a company's own plantation labor force, as the prices farmers receive depend on the quality of their output. Perhaps most important, contracting gives buyers more control over the production process than they have in open-market relations, as contracts usually specify the procedures to follow in cultivation and harvesting. To facilitate supervision and control of farmers, contracting agencies establish their own agricultural extension services, which enable them to provide more regular contact with farmers than is characteristic of typically underfunded government agricultural extension services. A study of contract farming in sub-Saharan Africa, for example, reveals that ratios of extension agents to farmers in many schemes commonly range from 1:50 to 1:200, whereas ratios of government agricultural extension officers to the general farming public are usually 1:1,000 to 1:2,000 and above (Watts et al. 1988). Such emphasis on extension work in contract farming facilitates the production of high-quality output.

Beyond the basic elements common to all contract-farming enterprises, considerable diversity is evident in the structure of enterprises and the nature of the agreements, which can be either written or verbal (Watts et al. 1988; Glover and Kusterer 1990; Carney 1992; Jaffee 1994; Watts 1994a). Similarly, the range in the number of farms included in projects is quite wide; reviews of cases in sub-Saharan Africa reveal enterprises with fewer than one hundred farms to those with nearly two hundred thousand (Ayako et al. 1989; Watts 1994a), with the largest generally having significant state involvement as well as external donor financing (Watts et al. 1988).

A wide range of crops are grown under contract farming, with most destined for export. Quantitatively, traditional export crops such as bananas, oil palm, tea, sugar, and rubber predominate, while "nontraditional exports" such as cauliflower, broccoli, asparagus, melons, and flowers are increasing the most rapidly (Minot 1986; Glover and Kusterer 1990; Watts 1994a).[2] Although any commodity can be produced in these schemes, certain crop characteristics are more likely to be found. Minot (1986) and Goldsmith (1985) note that such commodities are usually perishable, processed, and labor intensive; require considerable care in cultivation; have high value per unit volume or weight; and/or have economies of scale in certain phases of production. Although these characteristics are clearly important, research (Wilson 1986; Little and Watts 1994) has shown that whether contracting is employed is also strongly influenced by political-economic considerations; for example, state objectives, such as improving socioeconomic conditions in poor areas or facilitating resettlement, may take priority over purely technical and agronomic considerations in deciding whether contract farming is used (Jaffee 1994).

Although contract-farming schemes can include farms of almost any size, from peasant microplots to thousand-acre plantations, peasants figure prominently in such enterprises in developing countries (Watts et al. 1988).[3] Reasons for their involvement vary considerably, often according to whether capital or the state is the dominant force behind initiating and managing the system. Transnational corporations may prefer peasants because they are less powerful and more easily controlled than are large-scale landowners (Glover and Kusterer 1990: 134). Peasants can also produce under conditions that are unprofitable for larger enterprises for two reasons: they grow some of the food that they consume by also engaging in subsistence cultivation (the classic "subsidy" to capital) and thus do not have to rely exclusively on returns from contract production to survive; and they exhibit the well-known capacity for "self-exploitation" of household labor—the tendency to continue working in spite of low or declining commodity prices because of the need to guarantee household reproduction. Furthermore, such growers are more capable of mobilizing labor in the precise quantities needed on a short-term basis (Clapp 1988). And they may also be preferable when detailed, meticulous care of crops is required to produce the high quality needed, for they are more highly motivated and need less supervision than plantation workers (Watts et al. 1988). In contrast, states sometimes have different objectives. Where equity concerns, political patronage, expanding rural participation in commercial activities, and/or resettlement are high priorities, state-run enterprises will likely favor small- and medium-sized farms.

Indeed, state presence and influence in these schemes is considerable, with projects managed by statutory corporations and public-private ventures being widespread (Glover 1984; Little and Watts 1994). State motivations for fostering contract farming are a complex blend of political and economic interests. Governments, often backed by international donor agencies such as the World Bank, the Agency for International Development, and the Commonwealth Development Corporation, are receptive to contract-farming schemes because such schemes are supposed to improve farm productivity and incomes, facilitate technology transfer, and contribute to rural development.[4] Welfare considerations, particularly in cases targeting the peasantry, are also sometimes paramount (Little 1994). In addition, economic pressures to expand contract production have been considerable, especially in the contemporary era of burdensome foreign debt, structural adjustment programs, and mounting food import dependency (McMichael and Myhre 1991; Little and Watts 1994). States employ contract farming to boost production of traditional commodity exports and to contribute to import substitution programs. At the same time, many traditional commodity exports have been declining in value, affected by increased international competition, saturation of markets, and "substitutionism"—the development of manufactured substitutes for traditional tropical commodities, especially tropical oils and sugar (Goodman 1990; Friedmann 1991). Thus, many states have turned to "export substitution" by fostering the production of nontraditional commodities, such as specialty horticultural crops, off-season fruits and vegetables, and flowers—many of which are grown under contract and are destined for affluent consumers and niche markets in advanced capitalist countries (Glover and Kusterer 1990; Collins 1993; Llambi 1994; Raynolds 1994a,b; Watts 1994a).

As ideal as contract farming may appear, it can be fraught with problems. As Wilson (1986) and others (see Carney 1992; Clapp 1994; Watts 1994a) have indicated, the contractual relation is not based on equal bargaining power between the parties involved; rather, it is asymmetrical, with the contract serving to facilitate surplus extraction and labor control. The schemes are also breeding grounds for intense struggles, conflicting interpretations, manipulation, and dishonesty. Perhaps the most contentious aspect is the determination of grading standards, with contracts specifying how commodities are to be graded, the respective price to be paid for each grade, and the criteria for rejecting substandard output (Glover 1984; Carney 1992; Watts 1994a). Grading standards are often quite complex and open to subjective interpretations. Indeed, arbitrary and unfair grading practices are the most common complaints of contract farmers (Glover 1984). Some purchasers unfairly apply more stringent interpretations of standards and increase rejection rates at times of market oversupply to lower their procurement costs and inventories (ibid.; Carney 1992; Collins 1993). In some cases, rejection rates have been over 50 percent (Glover and Kusterer 1990), while in even more extreme situations, buyers have unilaterally terminated contracts, creating severe financial hardships for growers (ibid.; Carney 1992).

At the same time, growers are not without their own methods of evasion (Daddieh 1994; Jackson and Cheater 1994; Jaffee 1994; Watts 1994a). Diversion of agrochemical inputs obtained on credit to noncontracted crops is a widespread practice (Porter and Phillips-Howard 1997). Where purchasers do not have monopoly control over markets, farmers sometimes sell to alternative buyers, making it impossible for contractors to recoup expenses for credit and extension; this problem of "leakage" has, in fact, occasionally threatened the viability of enterprises (Jaffee 1994). Moreover, the extent to which farmers faithfully follow the procedures specified in contracts and the advice provided by extension officers is highly variable, a crucial issue that is especially relevant to this book.

Clearly, the study of contract farming involves a complex drama among peasants, capital, and the state and raises numerous issues of fundamental interest in the study of economic development. Perhaps most critical is the appropriate conceptual framework for understanding this institutional form. Two contemporary trends are evident; one is to analyze contract farming in relation to the literature on industrial restructuring in advanced capitalist countries, and the second is to focus on the globalization of agriculture. I want to examine both themes critically before emphasizing the significance of the political ecology perspective for the study of contract farming.

Contract Farming, Capital, and the Industrial Restructuring Literature

Many researchers interested in the role of capital in agrarian change and contract farming now turn for inspiration to the burgeoning, but controversial, literature on industrial restructuring in advanced capitalist countries (see Storper and Scott 1992; Amin 1994). Two perspectives in this literature—those on regulation theory (see Lipietz 1986, 1987, 1992) and flexible specialization (see Piore and Sabel 1984)—have had marked impacts on recent agrarian studies. A brief discussion of these perspectives as applied to the industrial sector is thus necessary before considering the controversies surrounding their application to the realm of agriculture in general and contract farming in particular. My concern here is not with the numerous differences between regulation theory and the flexible specialization approach (see Hirst and Zeitlin 1992) but with the industrial restructuring concepts that have been incorporated into analyses of agrarian change.

The literature on regulation theory emphasizes that understanding the evolution of capitalism and associated patterns of accumulation, crisis, and change requires consideration of the linkages among production and labor processes, consumption, and forms of institutional regulation. One of the central elements of this perspective is the periodization of capitalist development into different "regimes of accumulation." "Fordism," a regime of accumulation lasting from the 1940s to the early 1970s, is viewed as an industrial system based on mass production of standardized commodities linked to expanding mass consumption and regulated by Keynesian state policies, the welfare state, and collective bargaining (Lipietz 1986, 1987, 1992). Production incorporates a Taylorist division of labor, in which large numbers of semiskilled workers perform repetitive tasks on assembly lines. Wage increases tied to productivity gains created an expanding market for consumer goods, and competition among firms centered on price competition afforded by economies of scale. Fordism emerged first in the United States and subsequently in Western Europe, reaching its zenith in the post–World War II era.[5]

In the late 1960s and early 1970s, however, Fordism in North America and Western Europe began to face a series of severe internal and external crises—collapse of the Bretton Woods system of fixed exchange rates, increased competition from Japan and the newly industrializing countries (NICs) in East Asia and Latin America, the sharp rise in oil prices, stagflation, declines in the rate of productivity growth, rising unemployment, growing worker unrest, and decreasing corporate profitability (Schoenberger 1988; Ticknell and Peck 1992). The significance of these changes for Fordism and mass production has been the subject of much debate, but adherents of the flexible specialization approach have been particularly explicit concerning the impact of these crises of the 1970s.[6] They assert that in response to these pressures in an era of heightened international competition, some industries and services have adopted more "flexible" or "post-Fordist" patterns in the use of labor, technology, and resources, in corporate organization, and in inventory management. Examples of the trend toward flexibility include the use of computer-programmable machines to produce a range of product types (in contrast to the fixed-purposed machinery in mass production); adoption of just-in-time inventory management (in which parts are delivered just as they are needed to reduce inventory costs); and vertical disintegration (as opposed to vertical integration under Fordism) and the concomitant growth of subcontracting (Storper and Scott 1986; Schoenberger 1988). Instead of the mass production of standardized components for mass consumption characteristic of the Fordist era, firms began to shorten production runs, make more rapid changes in product design, and focus on differentiated consumer styles and niche markets (Ticknell and Peck 1992). In the context of labor relations, a variety of patterns emerged. Instead of a large number of workers performing a single task on assembly lines, with minimal responsibility, which occurred under Fordism, a core group of workers now saw their jobs encompass multiple tasks with more skill requirements and greater responsibility in decision making. Firms began emphasizing numerical flexibility in labor employment, rapidly adjusting quantities of labor used by complementing their permanent labor forces with temporary, low-paid, and low-skilled workers and by relying increasingly on subcontracting. In addition, although pricing continues to be important in terms of competition, increasing emphasis is placed on product quality, customization, and timeliness.

These changes have been significant in certain sectors of the economy (such as electronics). But whether they represent a modification of existing trends or herald a radical break with Fordism and mass production and indicate a new "post-Fordist" era of flexible specialization is the subject of intense debate on both empirical and theoretical grounds (see Amin and Robins 1990; Moulaert and Swyngendouw 1989; Sayer 1989; Gertler 1992; Ticknell and Peck 1992; Goodman and Watts 1994; Page and Walker 1994; Bonanno and Constance 1996; Ó hUallacháin 1996). The immediate concern here is not to descend into the mire of such debates but to examine the relevance of the literature on industrial restructuring for understanding the growth of contract farming. Indeed, numerous researchers have argued that such literature is particularly appropriate not only for understanding change in industry but also in agriculture (see Kenney et al. 1989, 1991; Sauer 1990; Watts 1990, 1992; Bonanno 1991; Goe and Kenney 1991; Kim and Curry 1993; Ufkes 1993; Cloke and Le Heron 1994; Raynolds 1994b). Two fundamental questions must be addressed. First, are the characteristics associated with Fordism and post-Fordist flexible specialization useful for understanding the growth of contract farming? Second, and more crucial, is it appropriate to apply concepts from the literature on industrial restructuring to the analysis of agriculture? My own view, which is parallel to that expressed by Goodman and Redclift (1991), is that processes described as being both Fordist and post-Fordist in the wider economy have influenced agrarian change but that application of such conceptual frameworks to the realm of agriculture has fundamental weaknesses (see also Goodman and Watts 1994; Page and Walker 1994; Labrianidis 1995). In the case of contract farming specifically, characteristics linked to both Fordism and post-Fordism are relevant to the growth of this institutional form, though the tendency in the literature has been to associate contract farming more with the flexibility of the alleged post-Fordist era.

Concerning the first issue, contract farming in developing countries has been well suited to supplying the requirements of Fordist food-processing industries in advanced capitalist countries, given the need for uniform, standardized commodity inputs and rigid scheduling of delivery times (Kenney et al. 1991). Rising wage levels in Fordist economies led to increased mass consumption not only of meats and durable (prepackaged and frozen) foods—the hallmark of the "Fordist diet" (Friedmann and McMichael 1989)—but also of imported fresh fruits and vegetables, including bananas. Contract farming in developing countries has links to Fordism besides its being a supplier of commodities. Contract growers are heavily dependent on the importation of Fordist-produced inputs, primarily agrochemicals and, to a much lesser extent, farm machinery. In addition, the spread of Fordist industrial production techniques to the NICs, while not stimulating the same level of consumption as in advanced capitalist countries (Lipietz 1986), has nevertheless led to growth in the number of people in middle- and upper-income groups, who, along with an expanding proletariat, increased domestic markets for fresh and processed foods, part of which have been supplied by such farmers (McMichael 1992).

Patterns associated with flexible specialization can also be viewed as stimulative to the growth of contract farming. Numerous researchers (Wilson 1986; McMichael 1992; Watts 1992, 1994a,b; Collins 1993; Raynolds et al. 1993; Raynolds 1994b) have argued that the flexibility characteristic of industry responding to heightened competition is also evident in contract farming. In particular, they assert that contract farming is a form of subcontracting that, as in industry, increases flexibility by limiting both capital's direct investment costs and labor management problems while increasing firms' abilities to make quick changes in response to new market conditions. Another parallel can be seen in increasing product differentiation (Goodman and Redclift 1991; Goe and Kenney 1991).[7] Watts (1994b) notes that contract farmers are employed to produce a variety of exotic crops and multiple varieties of the same commodities that were previously rare or unavailable in supermarkets in advanced capitalist countries; such unprecedented variety in available produce reflects ever more differentiated consumer segments and the growth of niche markets (though not the disappearance of mass markets for other commodities, such as bananas). Lastly, the growing emphasis on product quality generally in the competitive arena of post-Fordism can be linked to the heightened significance of quality in agricultural products, a factor also conducive to the growth of this institutional form because it is often employed to produce labor-intensive crops requiring meticulous care to ensure high-quality output (Watts 1996: 233).

While the literature on industrial restructuring has been useful in forcing researchers to examine agriculture in relation to the broader economy and not as an isolated sector, application of such conceptual frameworks to the domain of agriculture is problematic. In particular, agriculture has distinctive qualities that make such an endeavor highly questionable (see Mann 1990; Goodman and Redclift 1991; Baxter and Mann 1992; Goodman and Watts 1994, Page and Walker 1994).

Part of the problem stems from understanding just what is required to make agriculture "Fordist" (Goodman and Watts 1994). Some suggest that large-scale enterprises employing advanced technology, engaging in monocrop agriculture, producing mass quantities of commodities at uniform levels of quality for mass markets, increasing productivity, employing large numbers of unskilled workers, and consumption of Fordist inputs all combine to make agriculture in the United States, or at least certain segments of farming, Fordist (see Kenney et al. 1989, 1991; Kim and Curry 1993; Raynolds 1994b). But the similarities between industry and agriculture are more illusory than real. Page and Walker's (1994: 13) comments are particularly appropriate here:

> Putting a Ford tractor in the field does not put Ford's system of production into place. . . . The notion of a Fordist assembly line makes little sense in terms of on-farm production. Agricultural technology includes elements such as hybridization, fertilization and irrigation that pertain first of all to the manipulation of natural processes of plant (and animal) growth, and only secondarily to the labor process. And the labor process is stretched out by natural growth cycles in a way that cannot be compared with the time-space compression of sequential processing or assembly in a factory.

Thus, one key difference is that the labor process in agriculture cannot operate continuously in assembly line fashion as it does in Fordist industry, given the biological constraints of agriculture. Moreover, the often seasonal migrant wage labor force (really an element of flexibility!) in much of large-scale agriculture in advanced capitalist countries, although having low levels of skill and performing repetitive tasks, is hardly comparable with the unionized workers of Fordism enjoying increasing consumption levels in response to rising productivity (see also Goodman and Watts 1994).

Attempts to label certain characteristics of agriculture as "post-Fordist" are equally problematic. For example, Goe and Kenney (1991) consider polyculture as an example of post-Fordist flexibility, yet polyculture—the cultivation of a variety of crops in the same field—has been practiced by tribal and peasant cultivators for centuries. Also, caution is required in attributing exclusive significance to the flexibility issue. Although transnational corporations sometimes do employ contract farming to increase flexibility, many contract-farming schemes have considerable state involvement, and state interests may differ fundamentally from those of capital. Vertical disintegration has also occurred in agriculture, but often in response to political pressures and fears of expropriation and not always to provide greater flexibility to address rapid changes in market conditions. In addition, although contract farming is geared toward producing high-quality output and quality concerns may be greater today than in the past, quality issues were also significant in agricultural marketing during the heyday of Fordism as well, as was true in the banana market in the United Kingdom. Finally, issues of labor relations must be examined. Post-Fordist labor processes involve a core of skilled, multitask workers complemented by low-paid, low-skilled temporary workers. Certainly, low-paid, temporary, and seasonal workers have long been an essential component of agriculture in both advanced capitalist countries (Pugliese 1991) and in developing countries even before advent of Fordism, let alone post-Fordism. And if we do consider contract farming as a function of post-Fordist trends, do such growers represent the new, skilled, multitask workers or the low-skilled, poorly paid, often temporary workers? Or are they really the deskilled workers characteristic of industrial Fordism? The confusion stems ultimately from the inherently different nature of agriculture.

In what sense, then, is agriculture different from industry? Several perspectives have been offered. Mann and Dickenson's (1978) well-known hypothesis (see also Mann 1990; Baxter and Mann 1992), which has stimulated considerable controversy, is very relevant here. Concerned to explain the limited involvement of capitalist enterprises in agriculture compared with those in industry in advanced capitalist countries, they assert that the biological and environmental contexts of agriculture make it less appealing to capital. A key constraint to capitalist investment is the nonidentity of production time and labor time. Labor requirements in agriculture are usually seasonal, episodic, and subject to the vagaries of weather, whereas production itself is continuous as the crops grow. In contrast, in industry there is generally an identity of labor time and production time, because labor can be employed continuously, which is crucial for efficient assembly line production (Baxter and Mann 1992). Capital can extract surplus value only while labor is employed, and thus the average rate of profit in agriculture is lower than that in industry. They describe a host of other land-based and environmentally based constraints that also differentiate agriculture from industry, including highly seasonal labor requirements, inability to employ machinery year-round, perishability of products, and environmental risks and uncertainties. These constraints combine to prevent the utilization of continuous flow, assembly line production that is evident in Fordist industry (ibid.). Indeed, Mann (1990) asserts, such differences actually facilitate the use of contract farming, which is often employed in labor-intensive commodities that entail inefficient, noncontinuous use of labor time, making them unattractive for direct investment by capital.

Another perspective is offered by Goodman and Redclift (1991), who emphasize the contribution of agriculture to the Fordist regime of accumulation but assert that biological constraints to agricultural production (such as the nature of "photosynthesis, gestation, species diversity, and land space" [91]) lead to patterns of capital accumulation in the agro-food system that are different from those in industry. In the case of the agricultural system, capital invests not primarily in direct farming but outside the immediate production process in off-farm industries that supply inputs to growers, develop substitutes for agricultural commodities, and process foods.

Although the theoretical basis of their argument differs from that of Mann and Dickenson, both are concerned with the implications of environmental and biological constraints in agriculture for capitalist investment and accumulation. My concern, while also highlighting the importance of the environmental context of farming, is different. Specifically, I emphasize the significance of two implications of the *environmental rootedness* of agriculture for farming in general and contract farming in particular. First, the characteristics of the environment itself—including both the physical environment and the biological nature of agriculture—are *creative forces* that affect the ability of capital and the state to control the peasant labor process and help shape the relationship between local food production and contract production for export. Second, the *farming experience*, which forces farmers to cope with the inherent spatial and temporal variability of rainfall, temperatures, pests, soils, slopes, and crop varieties, imbues growers with a more flexible perspective on productive activities than is characteristic of deskilled factory workers; coping with such variations also produces interpersonal differences among farmers in their production practices and strategies, patterns of considerable importance in understanding both the ability of capital and the state to control the labor process in contract farming enterprises and the problem of pesticide misuse.

Contract Farming and Globalization

Two closely related perspectives on the political economy of agriculture known as the "new international division of labor in agriculture" and the "new internationalization (or globalization) of agriculture" also figure prominently in the study of contract farming (Rama 1985; Sanderson 1985a,b, 1986a,b; McMichael and Myhre 1991; Bonanno 1991; Constance and Heffernan 1991; McMichael 1991, 1992, 1993, 1994a,b,c; Raynolds et al. 1993; Ufkes 1993; Watts 1994a,b; Bonanno et al. 1994).[8] The overarching concept unifying these perspectives is that of "globalization"—"the worldwide integration of economic process and of space" (McMichael 1994a: 277). The literature on globalization reflects a diversity of viewpoints (see Amin and Thrift 1994; Mittelman ed. 1996), including conflicting opinions concerning when the process is supposed to have begun, but most researchers trace the dawn of globalization to the economic turmoil of the late 1960s and 1970s (Brecher and Costello 1994; Mittelman 1996a; Cox 1996; Gereffi 1996; compare Koc 1994). Certainly, globalization is manifested not only in agriculture but also in a wide range of phenomena, including the spatial reorganization of economic investments and production, the deregulation and spread of financial markets, the diffusion of identical consumer goods to distant locales, population migrations, and the extended reach of "Western culture" (Mittelman 1996a: 2).

With respect to agriculture, several related themes are evident in this literature: Agriculture is increasingly integrated into the global economy and affected by such supranational forces as transnational capital, international financial institutions, including the World Bank and the International Monetary Fund, and international trade agreements and institutions, particularly the General Agreement on Tariffs and Trade (GATT) and its successor, the World Trade Organization (WTO)—all of which are aligned to accelerate the trend toward trade liberalization. These same influences are weakening the power of nation-states to regulate patterns of domestic accumulation and change in agriculture within their own territories, forcing them, in the words of McMichael (1994b: 5), "to pursue national competitiveness rather than national coherence with respect to their agricultural sectors."[9] The debt crisis has increased the regulatory power of multilateral financial institutions over nation-states. Transnational corporations are utilizing increasingly complex patterns of global sourcing—obtaining the same commodities from numerous locations scattered throughout the world—to reduce political and environmental risks and uncertainties related to procurement of supplies and to play one location against the others to obtain subsidies and other concessions, all of which ultimately help reduce their costs while contributing to the weakening of the regulatory power of nation-states. Also, internationalization of agriculture is causing a decline in domestic food production, and agriculture is becoming less a distinct sector and more thoroughly linked to industry as both consumer of industrial products and supplier of industrial inputs as part of agro-food complexes.

According to this view, a new international division of labor in agriculture has emerged, parallel to that in manufacturing (see Fröbel Heinrichs, and Kreye 1980; Sanderson 1986b; McMichael 1994c). In the old international division of labor, developing countries exported such classic crops as sugar, bananas, coffee, and coconuts and imported industrial products. While they continue to export such commodities, a new international division of labor in agriculture has emerged in which they also export a variety of high-value, labor-intensive, off-season, and nontraditional agricultural products and import low-cost basic grains and oil seeds from North America and Western Europe (Sanderson 1986b; McMichael and Myhre 1991; Constance and Heffernan 1991; McMichael 1994c). Similar to the pattern in industry in which segments of manufacturing moved from core to periphery in search of cheap labor (Fröbel, Heinrichs, and Kreye 1980), many of these "new" commodities, such as beef, poultry, flowers, broccoli, cauliflower, and tomatoes, that have traditionally been produced in core countries are now being produced in part in developing countries because of lower labor costs and to take advantage of different growing seasons (Constance and Heffernan 1991; Watts 1994a; McMichael 1994c).

Thus poultry-producing contract farmers in Thailand have become competitors with the United States as suppliers of chicken to Japan (McMichael 1993), and Mexican contract farmers now raise feeder calves for the United States, which previously produced them internally (Sanderson 1986a,b).

But what is significant about the new international division of labor in agriculture and the new internationalization of agriculture, according to Sanderson (1986a,b), is not the change in trading patterns per se or the types of commodities produced but the changes in production relations. Regions are no longer linked only through trade but also through production and investment, and transnational capital is now the dominant force shaping the trajectory of agriculture. Most important, transnational capital sets standards for production technologies and for quality and grading criteria that have become increasingly influential not only in the international realm but also in the domestic arena; thus even national capital producing solely for their domestic markets in developing countries is still affected by such internationally accepted standards (Sanderson 1986b). Standardization facilitates the policy of global sourcing, as uniformly produced products can be obtained from throughout the world (Constance and Heffernan 1991; Ufkes 1993). Although transnationals continue to rely, in part, on their subsidiaries throughout the world for agricultural commodities, an increasingly important dimension of global sourcing is the utilization of contract farming, which employs standardized technologies and grading criteria.

The literature on the new international division of labor in agriculture and the new internationalization of agriculture forces us to broaden our horizons beyond the farm gate (Whatmore 1995). It emphasizes the global, supranational forces impinging on both nation-states and their systems of agriculture. As this literature suggests, the ability of nation-states to determine their own domestic economic policies independently has been hampered by the increased mobility of transnational capital, the growing power of multilateral financial institutions (Bonanno et al. 1994; Constance, Bonanno, and Heffernan 1995), and the creation of supranational trading blocks, such as the EU and the North American Free Trade Agreement (NAFTA). The ascendancy of the EU and its Single European Market (SEM) has been especially important in relation to the banana industry of the Windward Islands, because it has severely curtailed the United Kingdom's ability to continue providing traditional market protection for bananas imported from its former colonies, a constraint that is now threatening the very survival of the Windwards banana industry. Furthermore, trade liberalizing influences emanating from GATT and the WTO have contributed to pressures to reduce protection for Windwards bananas.

As illuminating as this literature has been, certain problems are evident. One weakness in some of it is the firm conviction that an era of global regulation has replaced national regulation, with transnational capital and trade liberalization being declared victorious over the nation-state (for example, Bonanno 1991; McMichael and Myhre 1991; compare Bonanno et al. 1994). The impression gained—that states previously were able to determine national patterns of accumulation effectively—is questionable (Panitch 1996: 84–85), especially in the case of the small islands of the Eastern Caribbean, including the Windward Islands, whose economies have long been subject to control by foreign states and capital (see Richardson 1992). Moreover, state policy is by no means insignificant in mediating such global forces (Raynolds 1994a; Goodman and Watts 1994; Llambi 1994; Panitch 1996), though the degree to which states can cope with and even shape the process of globalization is highly variable (Raynolds et al. 1993: 1104).

That domestic food production has declined in many developing countries is clear. But much of the literature seems to portray such declines as an inevitable result of the forces of globalization (see Barkin 1985; Sanderson 1986b; Rama 1985; McMichael 1992; Raynolds et al. 1993; Bonanno 1992, 1994). The literature glosses over the diversity of local responses as well as the precise mechanisms involved in such declines. In essence, there is often very minimal understanding of the dynamics of agricultural systems in much of this literature, but analyzing the historical contexts and intricacies of agricultural change is absolutely crucial for illuminating the causes of declines in local food production. While Goodman and Watts (1994: 38, 43) want to know what is "new" about the new internationalization of agriculture literature, we also need to pose the question of what about it is "agricultural"! As I show in this book, the decline of domestic food production is a much more complicated affair than just a residual impact of the grand sweep of globalization.

Other weaknesses are also evident. The literature focuses primarily on patterns reflected in new exports of off-season and nontraditional fruits and vegetables, meats, and flowers, while indicating limited interest in such traditional commodity exports as bananas, which are also subject to these same globalizing influences. Moreover, such research tends to overemphasize the extent to which production technologies have been homogenized. Local historical, social, and environmental conditions, as well as strategies by transnational capital, can lead to the development and utilization of nonstandardized technologies geared toward producing commodities that meet standardized expectations of quality grading, as has happened in the Windwards banana industry.

For those following what Terence McGee (1978: 101) calls "the dirty boots begets wisdom" approach—that is, relying on detailed, local-level studies out in the field in developing countries for the collection of primary information to inform our analyses, as I utilize in this book—the literature also suffers from an overemphasis on structural forces (see also Whatmore 1995; Jarosz 1996a). The world of real people, strategies, perceptions, conflicts, household relationships, and cultural histories all become homogenized before the advance of the forces of globalization. The critique of Lowe, Marsden, and Whatmore of the related literature on food regimes is particularly applicable: "[T]here is a risk of imposing a categorical accumulationist logic that diminishes the significance of social agency, regional diversity and contingency, and that thereby fails to grasp the differential integration of agriculture into the global economy" (1994: 9; see also Mittelman 1996b: 232).

Thus, we cannot view the patterns associated with contract farming simply as a reflection of the forces of globalization. An approach is required that will allow us to appreciate more fully the importance of diversity, agency, and local context, while incorporating the significance of broader structural forces. Moreover, such an approach must also be sensitive to the "environmental rootedness" of agriculture. Political ecology is particularly suited to the task.

The Political Ecology Framework

Political ecology, which is becoming increasingly popular, is an especially useful framework for analyzing human-environment relations in agriculture in general and contract farming in particular.[10] It combines perspectives from cultural ecology, which have been prominent in geography and anthropology (see Hardesty 1977; Porter 1978; Netting 1986; Butzer 1989), and political economy. Traditional cultural-ecological studies explored the intricate, complex interactions between peoples and their environments in the context of resource use. Usually focused at the microscale, several themes captured the attention of cultural ecologists. One linked patterns of livelihood and sociocultural institutions to the process of adaptation to the natural environment. Another examined the manner in which behavior and beliefs functioned to regulate human-environment relations.[11] But such analyses failed to consider the role of political economy, which focuses on the nature and significance of the unequal distribution of power and wealth in society. Amalgamating the two perspectives, political ecology emphasizes that human-environment relations at local, regional, and global scales can be understood only by examining the relationship of patterns of resource use to political-economic forces. Thus, it is crucial to investigate how agriculture and environmental change are influenced by state policy, regional trading blocks such as the EU, investments by transnational capital, penetration of the market, and the social relations of production (patterns of control over land, labor, and surplus) (Grossman 1984a; Blaikie and Brookfield 1987a; Schmink and Wood 1987; Bassett 1988a, 1994; Carriere 1991; Peet and Watts 1996). Political-ecological studies also highlight the need to understand such processes in their historical contexts, because as Blaikie (1995a: 13) asserts, the "complex interactions between environment and society are always embedded in history and locally specific ecologies." In particular, the historical development of unequal patterns of control over land, labor, and surplus greatly influences current resource use. At the same time, the intricate analyses of local-level, human-environment relations characteristic of studies in cultural ecology must remain integral to political ecology. Thus, the implications of crop ecology, changing patterns of intercropping, soil management, livestock herding, and new forms of technology are just as fundamental to political ecologists as they are to cultural ecologists. Similar to cultural ecology, political ecology has focused primarily on developing countries, though exceptions exist (for example, Black 1990).

Blaikie and Brookfield's (1987a) widely cited work, in which they developed their conceptual framework (see also Blaikie 1985) for "regional political ecology," was the propulsive force driving a plethora of studies using the term "political ecology" in the late 1980s and 1990s.[12] But studies using the essence of a political ecology approach, though not employing the term, were already of growing importance in the 1970s and 1980s. An arena of particular interest for such studies was the impact of colonial policies on human-environment relations (see Kjekshus 1977; Porter 1979; Franke and Chasin 1980; Watts 1983). Other interests included the role of the state in deforestation (Hecht 1985), the significance of economic boom and bust cycles (Grossman 1984a), and the green revolution (Yapa 1979). What researchers now call political ecology also has similarities to what I (Grossman 1984a) called "the cultural ecology of economic development" and Yapa (1979) termed "ecopolitical economy." At the same time, today's studies have a more sophisticated treatment of political-economic concepts, such as the state, capital, and class, than did these previous works.

For most political ecologists, the focus on human-environment relations is at the local level—examining resource-use patterns as they relate to households as units of production, struggles (especially gender-based) within households, or relations/conflicts among households. This priority of highlighting local-level, human-environment relations reflects, in part, the heritage of cultural ecology. Furthering this trend, many follow Blaikie and Brookfield's (1987b: 27) strategy of focusing initially at this level:

> It starts with the land managers and their direct relations with the land (crop rotations, fuelwood use, stocking densities, capital investments and so on).

> Then the next link concerns their relations with each other, other land users, and groups in the wider society who affect them in any way, which in turn determines land management. The state and the world economy constitute the last links in the chain.

The initial focus in political ecology highlighted the relations among state policy, surplus extraction, accumulation, and environmental degradation (for example, Watts 1983; Blaikie 1985; Hecht 1985; Blaikie and Brookfield 1987a; Collins 1987; Bassett 1988a, 1994; Hershkovitz 1993; Stonich 1993; Zimmerer 1993). These studies provided a necessary corrective to viewpoints asserting that environmental problems are related primarily to Malthusian pressures, peasant irrationality, and ignorance. Political ecologists have also begun exploring a wider range of issues relevant to understanding patterns of resource use, including the role of social movements, gender conflicts, the persistence of the peasantry, and environmental conservation programs (Zimmerer 1991, 1996a; Carney 1996; Jarosz 1996b; Leach and Mearns 1996; Moore 1996; Peet and Watts 1996; Schroeder and Suryanata 1996). Part of this broadening agenda is an emphasis on understanding social constructions of the environment. Researchers stress that our views and discourses about environmental problems are social constructions reflecting our own backgrounds, values, and positions of power, not absolute truths; such multiple constructions and associated environmental discourses frame differing, contested points of view and suggested solutions (see also Zimmerer 1996a; Bryant 1996; Moore 1996; Jarosz 1996b). Given the concern with political economy, political ecologists are thus examining how differences in the economic and political power of social groups, defined on the basis of such characteristics as occupation, class, gender, and age, affect variations in perceptions, interpretations, and environmental discourses—which reflects the growing poststructural concern with discourse in the social sciences generally (Leach and Mearns 1996; Peet and Watts eds. 1996).[13] For example, Zimmerer (1996a: 112), in his analysis of soil erosion in Colombia, stresses the importance of a "focus on discourses representing the ideas and ideologies held by groups of individuals and institutions." Such analyses have certainly contributed to our understanding of human-environment relations. But we also need to examine more critically variations in environmental perceptions, interpretations, and beliefs at the individual level—that is, interpersonal differences within such groups. And such crucial interpersonal differences—which are a function of the environmental rootedness of agriculture—have significant implications for patterns of resource use.

A key issue in political ecology has been the environment itself. The major concern has been the appropriate conceptual framework for characterizing the process of environmental change and the comparative utility of such ecological concepts as "homeostasis," "resilience," and "disequilibria" in human-environment analyses (for example, Blaikie and Brookfield 1987a; Zimmerer 1994). My own interest, however, focuses on the role of the environment in our explanations (Grossman 1993). Some studies focus only on the effects of political-economic forces on the environment in a unilinear fashion (for example, Schmink and Wood 1987). Indeed, in a review of the field, Zimmerer (1996b: 179) complains that "most political ecology has conceived the environment solely as a receptor of modification." While it is important to consider such influences on the environment, it is also crucial to analyze the significance of environmental variables themselves and how they interact with political-economic forces to affect human-environment relations (Grossman 1993). In essence, the environment, which is much more than a malleable entity molded by human activity, has significance for understanding a range of issues in political ecology, including control by capital and the state over the labor process in farming, the dynamics of agricultural change, and the problem of pesticide misuse.

A related issue concerns a classic problem in the social science literature in general—the structure-agency debate. That is, what is the relationship between impersonal, structural forces and the lives of living individuals at the local level, who have their own thoughts, aspirations, and beliefs? Some have expressed concern that political ecology allocates too much explanatory power to structural forces, portraying what happens at the local level simply as reflecting all-powerful, political-economic forces, while downplaying the crucial role of local-level resistance, strategy, intra- and interhousehold struggles over access to resources, and cultural beliefs and interpretations (Zimmerer 1991; Stonich 1993; Moore 1996). Thus, political ecologists are, appropriately, becoming increasingly sensitive to this issue, considering how such local-level forces interact with and mutually condition broader, structural forces (for example, Bassett 1988a; Zimmerer 1991; Stonich 1993; Blaikie 1995a,b; Moore 1996). This book indicates that a key dimension of human agency is related to the environmental rootedness of agriculture and its significance for the farming experience.

The perspective of political ecology is highly relevant to the study of contract farming. A key aspect of this institutional form is the penetration of the production process by capital and the state, and it is the production process that links peoples and their environments. In relation to the three themes at the heart of this study, political ecology directs our attention to crucial issues of widespread significance in the development literature. The labor question involves the issue of capital and the state's control over labor and the deskilling of labor, central themes in the literature on political economy. It demands a consideration of whether the environmental and social contexts of farming at the local level make generalizations about the relation of capital to labor developed in the study of industry applicable to contract farming. Examination of the relation between local food production and contract farming for export requires a detailed analysis of the intricacies of agricultural systems, the environmental contexts of production, and the influence of capital and the state on inequalities in control over land, institutional subsidies, and the nature of market forces. Finally, understanding the environmental consequences of contract farming necessitates a consideration of both the political-economic forces molding agrochemical use and the multiple perceptions concerning pesticides at the local level.

The Labor Question

Contract farming has substantive implications for the peasant labor process. Capital and the state intervene directly in production, raising the contentious issue of labor control and its characterization and significance. Many contracts are quite detailed in outlining the procedures for and scheduling of planting, cultivation, and harvesting. Capital and the state attempt to achieve compliance with such contractual obligations through the provision of extension supervision and specification of quality standards.

The issue of control, which bears directly on how one views the relative significance of structural forces and local-level responses, is central to the debate concerning the characterization of contract farmers. Some researchers (Davis 1980; Clapp 1988, 1994; Watts 1994a) view them as "disguised" or "concealed" wage laborers or "propertied proletarians," in essence being more comparable with wage laborers than with independent farmers.[14] For example, Clapp (1994: 81) asserts that "contract farming is a form of disguised proletarianization: it secures the farmer's land and labor, while leaving him with formal title to both. . . . [T]he farmer's control is legal but illusory." In essence, such a view portrays contract farmers as "hired hands on their own land" (Little and Watts 1994: 16). That contract farming gives capital and the state some degree of control over the production process is clear. What is at issue is whether that control is so thorough and dominating that contract farmers' autonomy is reduced to a level comparable with that of wageworkers.

Concepts such as disguised wage labor are inappropriate for an understanding of the position of labor in these schemes for several reasons.[15] Contract farmers retain control over their land, a significant contrast to "free" wage laborers (Boesen and Mohele 1979; Buch-Hansen and Marcussen 1982; Mann 1990; Baxter and Mann 1992). Such control is not a mere formality but an essential element in the resilience and survival of peasant households. By retaining control over land, these farmers have an "exit option" (Hyden 1980)—they can terminate their contractual relations and switch to alternative land uses. Although tenants on state-run, contract-farming schemes are vulnerable to eviction for failure to follow specified procedures, most other nontenant contract farmers—who make up the large majority of contract farmers in developing countries—do not face such risks. Failure to follow directives may mean ejection from schemes, but not dispossession of land; indeed, the ability of private capital to foreclose on peasant holdings for failure to follow directives or repay credit is severely limited (Watts et al. 1988). Depending on the situation of the growers (see Grossman 1984a), falling prices for contracted commodities can lead to a mass exodus from contracting.

Contract farmers, in reality, also retain a measure of control over the commodities they produce, irrespective of what individual contracts may specify. Such de facto control sometimes leads to the problem of "leakage"—farmers selling their products to buyers other than the contractor to escape credit repayments—a pattern that can diminish profitability for capital and the state involved in such enterprises. Proletarians, in contrast, do not control the commodities they produce (Mann 1990).[16]

Furthermore, the concept of "disguised proletarian" inappropriately treats contract farmers as an undifferentiated mass. In reality, economic differentiation is often associated with contracting (Buch-Hansen and Marcussen 1982; Korovkin 1992). Some contract farmers reinvest earnings to expand their landholdings, increase expenditures on inputs, and employ more wage labor, which facilitates further accumulation—a process alien to the wage labor form.

The concept also diverts attention from another dimension not characteristic of wage laborers—the need to organize a labor force, which can have major ramifications. Contract growers marshal the labor of other family members, relatives, and friends in complex networks. Indeed, Lehmann (1986) has asserted that peasant access to such low-cost social networks, which are unavailable to larger capitalist farms, is a key factor facilitating peasant survival. Organization of household labor, however, is not unproblematic. The well-known capacity of peasant households for self-exploitation of labor, while generally considered a virtue in the literature because it helps ensure the competitiveness of peasant agriculture, can lead to significant intrahousehold gender struggles. Carney (1988, 1994), for example, has demonstrated that the intensification of labor associated with contract farming can lead to gender conflicts within a household in relation to control over labor and land, which can ultimately limit the ability of capital and the state to control the peasant labor process, extract surplus, and accumulate capital. Similarly, von Bülow and Sørensen (1993) reported that in their survey of Kenyan contract tea farmers, one-third of the tea plots were partly or completely neglected mainly because of household labor problems linked to conflicts between spouses.

Ultimately, the concept of disguised wage labor is intended to portray the clear dominance of capital and the state over labor and the extremely limited autonomy of contract farmers—in a situation parallel to wage laborers who do not control any means of production. Clearly, the degree of control exercised by capital and the state is highly variable, not unproblematic, in contract farming (Llambi 1990; Jaffee 1994; Little 1994). Contracts vary greatly in the degree of specificity of tasks to be performed and, more important, in the extent to which contract farmers adhere to such regulations (see Glover and Kusterer 1990: 127; Porter and Phillips-Howard 1996: 291). Some schemes are rigidly regulated, as in the well-known Kenyan cases of sugar production in the Mumias scheme and tobacco production under British American Tobacco (BAT) (see Buch-Hansen and Marcussen 1982; Shipton 1985). The banana industry in the Caribbean is characterized by significantly more autonomy for its farmers. We must view the degree of control over peasant labor in contract farming not as a foregone conclusion but as an arena for investigation (Little 1994). Contracts are sites of struggle, not acquiescence.

Another dimension of the "labor question" that also reflects the problems of autonomy and control concerns whether contract farming leads to a "deskilling" of labor. The debate has its origins in the Marxist literature on the capitalist labor process and centers on the work of Braverman (1974). In his well-known *Labor and Monopoly Capital,* he asserts that an inherent tendency of the capitalist mode of production is the deskilling of labor, which gives capital greater control over labor and increases profitability.

In considering the decline in skills, Braverman employs an idealized conception of the nineteenth-century craft worker, whom he portrayed as having a wide range of knowledge and skills; high levels of individual discretion, autonomy, and control; and the ability to perform many of the tasks contributing to the manufacture of a product. In contrast, in the twentieth century, capital's utilization of Taylorist management strategies led to a deskilling of labor, which has three interrelated dimensions (Wood 1987). First, there is a separation of conception and execution. According to Braverman, management monopolizes knowledge about production and alone is involved in the planning and design of work, whereas laborers are relegated simply to carrying out its directives. Second, the division of labor is increased as management subdivides the labor process. Fragmentation of the labor process is accomplished by utilizing mechanization and making workers responsible for just one small part of the overall production process. It reduces dependence on costly skilled laborers, as management can subdivide work into segments so that the least-skilled laborer could be employed for each specific task. Third, as capital is the repository of knowledge and planning, it effectively controls labor, dictating the nature and pace of workers and thus reducing the autonomy of labor.

Braverman's book has led to an explosion of research on the topic, receiving both support and a "veritable avalanche of criticism" (Grint 1991: 190) from both Marxists and non-Marxists alike (see Wood 1982, 1987; Attewell 1987; Storper and Walker 1989). Problems noted by critics include a romanticization of the nineteenth-century craft worker (Wood 1982); portrayal of capital as "omniscient and able to achieve total control" (Wood 1987: 4) while ignoring the significance of worker resistance (Lee 1982) and the manner in which workers' own activities and interactions in the labor process create consent with respect to the rules of the workplace (Burawoy 1979); the characterizing of knowledge in a zero-sum format, with capital's accumulation of knowledge believed to be made at the expense of labor (Wood and Kelly 1982); and the assumption that capital's primary objective is control over labor, whereas its main goal is achieving greater profits, an outcome that can be accomplished in a variety of capital-labor formats (Wood and Kelly 1982; Grint 1991).[17] Most discussions of the subject have focused on the industrial sector, with some consideration also given to services in advanced capitalist countries.

Only recently has discussion of Braverman's work entered the agricultural realm, specifically in the context of contract farming. Several researchers have asserted that the concept of deskilling is also appropriate in contract farming (Clapp 1988, 1994; Mann 1990; Watts 1992, 1994a; Collins 1993; Baxter and Mann 1992; Cook 1994), a perspective that has gone unchallenged. But given the problem of applying concepts developed in the industrial arena to the analysis of agriculture, caution is warranted in uncritically adopting the concept. My analysis of banana contract farming suggests that the concept oversimplifies the nature of the labor process in contract farming.

First, there is the issue of separation of conception and execution. Clearly, most technologies and methods for the planting, cultivation, and harvesting of crops in contract farming are initially developed by the contracting agencies, not the farmers. But to assume that farmers do not use their own vast experiences developed in their specific environmental contexts is misguided. The juggling act that is farming, a function of the environmental rootedness of the enterprise, requires a melding of conception and execution. Farmers must make judgments concerning how to produce and deliver commodities that meet rigid quality specifications, taking into consideration their availability of labor, land,

and capital as well as their environmental constraints. As Kaplinsky (1988) notes in his discussion of labor processes in advanced capitalist countries, quality control, by its very nature, requires decision making.

Braverman's notion of the separation of conception and execution also is inadequate when considering peasant labor organization and recruitment. Crompton and Reid's (1982) critique of Braverman is relevant here. They assert that he emphasized only the loss of the craft element in deskilling, but many nineteenth-century craft workers also employed subordinate workers; deskilling meant a loss of this entrepreneurial function as well. For contract farmers, the management of the labor of others remains a primary concern that is, in addition, an element of conception in the labor process. Moreover, contract farming is not static. The introduction of new forms of technology requires contract farmers to creatively manage, organize, and readjust their recruitment of the labor of other household members, relatives, friends, and, at times, wageworkers, as will be revealed in this study.

Furthermore, to assume a clear delineation between conception and execution ignores the fact that most contract farmers produce other crops. They must carefully integrate contracted and other crops spatially and temporally into the agriculture cycle. Moreover, technological transfer may occur between contracted and noncontracted crops—such as in the use of agrochemical inputs—that may enhance farming skills overall (see Watts et al. 1988). Indeed, the relation among skills, technology transfer, and conception becomes even more complex as farmers, on their own initiative, sometimes misuse pesticides intended for contracted commodities by applying them on other crops, a problem that the concept of deskilling does not adequately cover.

The second dimension of deskilling—the subdivision of production into increasingly simplified stages—can occur in some cases. In contract production of sugar cane in the Mumias scheme in western Kenya, for example, growers in this highly regimented scheme perform only part of the labor involved in crop production, with the contractor providing work crews and machines to perform several tasks (Buch-Hansen and Marcussen 1982). But such a pattern represents only one extreme in contract farming. Indeed, as has been noted in general critiques of Braverman—what is crucial for capital is not the issue of control but profitability, and in contract farming, increasing profitability for capital can be obtained in a variety of ways. Segmentation of the labor process is not the only route. Indeed, in the case of the Windward Islands banana industry, the exact opposite pattern has occurred—the production process has become more complex over time as tasks once performed off-farm have been increasingly incorporated into farming—which has increased the skill content of the labor process to the benefit of capital.

The third dimension of deskilling—that of control by capital—has already been partly considered in the discussion of the concept of "disguised wage labor." That farmers have lost some autonomy is obvious, as they must produce a crop that meets certain quality and grading specifications; and producing to meet specifications, as this study will indicate, can have substantial implications for resource-use patterns in general. But control over peasant labor is certainly less thorough than in industrial and service sectors. Indeed, if control were effective, one would expect a uniformity in production practices, with farmers following all contract directives. Although contracts may be very specific in detailing the manner and exact timing of planting, cultivation, and harvesting, the degree to which farmers uniformly follow such directives is highly variable (Jaffee 1994; Little 1994). The widespread diversity of production practices, in turn, has significant implications for human-environment relations. Moreover, crop characteristics themselves may make control over the timing of activities difficult to achieve.

The Food Question

One of the most pervasive problems in developing countries is burgeoning food imports accompanying declining local food production. In relation to these patterns, an extensive literature emphasizes two related and overlapping themes—the impact of commodity production on household subsistence production and the impact of export agriculture on national food security (see Nietschmann 1973; Porter 1979; Watts 1983; Grossman 1984a; von Braun and Kennedy 1986).[18] Extending these lines of inquiry, researchers are increasingly concerned with the effects of contract farming on national food security (Glover 1984; Goldsmith 1985; Glover and Kusterer 1990; Watts 1990; Little 1994). Contract farming is of special concern for two reasons. First, it is usually very labor intensive, with some contracted crops requiring over two hundred days of labor per acre per year (Watts 1994a: 45); such labor intensity was captured by Shipton (1985: 295), who reported that "Luo farmers who have grown tobacco for BAT [under contract] say that they have never worked so hard in their lives." Second, capital and the state intervene in the peasant production process in novel ways to shape the allocation of land and labor, which can undermine local food production.

Discussions about the impacts of contract farming can be easily muddled because of the diversity of contractual contexts. Thus, while most contracted crops are destined for export, others play an important role in expanding traditional local food production, as does rice growing in The Gambia (Carney 1988), and in contributing to import substitution, as with contract oil seed and sugar production in Kenya. Contract farming can also have diverse impacts at the farm level. Some contracts limit the amount of land that can be devoted to contracted crops in order to regulate supplies and ensure adequate household labor availability for production (Buch-Hansen and Marcussen 1982). Conversely, some schemes prohibit intercropping of food crops with contracted commodities to maximize contract production (Daddieh 1994).

Similarly, other factors complicate analyses, indicating that the relationship is more complex than earlier critiques (such as those by Lappé and Collins 1977 and Dinham and Hines 1984) indicated. Contract production may be replacing other forms of commodity production, not subsistence production (Glover 1984). Crops rejected as substandard by contractors may augment local food supplies (Glover and Kusterer 1990). And inputs intended for contracted crops are sometimes diverted to the benefit of local food production.

Nonetheless, contract farming for export can have significant, negative impacts on food production (Bassett 1988b; Mbilinyi 1988; Glover and Kusterer 1990). The precise mechanisms leading to such results vary but usually revolve around conflicts in the productive sphere that undermine local food production. Specifically, the allocation of land and labor to contract farming, which is influenced by capital and the state, can conflict with and detract from the allocation of these inputs into food production. Contract production may be either very demanding of labor throughout the year or require significant labor inputs at certain seasons of the year, which creates seasonal labor bottlenecks that limit the time available for domestic food production (Bassett 1988b; Watts et al. 1988). Similarly, the areal expansion of contract farming can reduce land available for food production or push food gardens farther away from residences into more distant and often more marginal and erosion-prone lands. Such impacts are rarely felt uniformly within a community, however, as those with greater resources and wealth can cope with such constraints in a variety of ways, including the employment of wage labor.

Researchers focusing on the Caribbean tend to blame banana production in the Windward Islands for declines in domestic food production and for growing dependency on food imports (Marie 1979; Long 1982; Rojas 1984; Axline 1986; Thomson 1987; Thomas 1988; Nurse and Sandiford 1995). But in this book, using a political-ecological approach, I reveal a much more complicated picture. Part of the problem stems from the failure of previous researchers to analyze the dynamics of Windwards agricultural systems carefully. In particular, the environmental contexts of production, crop characteristics, the nature of cultivation techniques, and inequalities in control over land interact to encourage production of both local food crops and contracted bananas grown for export.

Moreover, we also need to look beyond patterns within St. Vincent to understand the food problem. The literature on the new international division of labor in agriculture and the new internationalization of agriculture, as well as that dealing with the related perspective of food regimes (see Sanderson 1985b, 1986b; Friedmann and McMichael 1989; McMichael 1991; Bonanno 1992; Raynolds et al. 1993), are very relevant here. Such research indicates that we need to consider patterns of agrarian change in advanced capitalist countries to understand food problems in developing countries. According to this literature, the geopolitical and economic policies of the United States and Western Europe resulted in the provision of generous agricultural subsidies to expand grain production, leading to increased exports through trade and aid. The resulting expansion of supplies of low-cost grain on the world market available for food imports has, in turn, depressed local food prices and production in developing countries (Friedmann and McMichael 1989; McMichael and Kim 1994). Such processes can negatively affect domestic food production independently of contract farming.

The Environmental Question

One of the hallmarks of contract farming is the intensive use of agrochemicals (see Murray and Hoppin 1992). Farmers apply pesticides and synthetic fertilizers with the expectation of reducing labor costs, improving quality, and increasing productivity. Dependence on agrochemical inputs has significant implications for human-environment relations. Pesticides can degrade both human health and the environment, lead to pesticide resistance, leave harmful residues on crops, and trap farmers in a financially binding and environmentally damaging "pesticide treadmill" (Thrupp 1994). Synthetic fertilizers can lead to reductions in fallow periods, lessen dependence on traditional methods of soil fertility enhancement, and change soil chemistry. Although chemical dependence in agriculture is increasing in general in developing countries since the 1950s, what makes the situation so unique in contract farming is the attempt by capital and the state to regulate the nature, extent, and timing of the application of agrochemical inputs. Contracting agencies foster intensive agrochemical use to obtain commodities that satisfy the demanding quality specifications and grading criteria of the market. At the same time, as the previous discussion of the labor process indicated, we should not assume that capital and the state are thoroughly effective in determining patterns of agrochemical use.

My concern in relation to the environmental question is the issue of pesticide misuse, the most serious environmental problem associated with contract farming. Political ecologists are increasingly emphasizing that interpretations of environmental problems are social constructions that reflect positions of power and biases, a perspective that has been fruitfully employed in analyses of deforestation (Jarosz 1996b) and soil erosion (Blaikie 1985; Blaikie and Brookfield 1987a; Leach and Mearns 1996; Zimmerer 1996a). We can extend this line of thinking to an examination of studies of pesticide use, which reveal similar biases in some interpretations by characterizing peasant pesticide-use patterns as "careless" and "indiscriminate." Such labels appear repeatedly in the literature and are similar to others, such as "irrational," "inappropriate technology," and "unproductive" that can be found in publications and official documents in relation to peasant involvement in deforestation and soil erosion. They are all employed to blame farmers and mask complex forces that contribute to environmental problems.

A political-ecological framework illuminates our understanding of pesticide misuse. One dimension of this approach is the specification and analysis of the political-economic forces affecting pesticide use. Certainly, regulation by capital and the state involved in contract farming is significant, but other pressures are evident as well. Concentration in the food retailing sector has led to the growth of large supermarket chains—which are particularly demanding of high-quality produce—at the expense of small, neighborhood grocery stores in advanced capitalist countries (Goodman and Redclift 1991; Nurse and Sandiford 1995); many contract farmers use pesticides to satisfy such intensifying demands for better-quality produce. Foreign aid programs often incorporate support for intensive chemical use. Also, the growth of export agriculture in general is associated with greater reliance on such inputs. And the role of transnational corporations in marketing pesticides in developing countries—sometimes unethically—is well documented (Weir and Schapiro 1981).

At the same time, what happens at the local level is not simply a reflection of broader, structural forces. The analysis of problems such as pesticide misuse must also delve into the dynamics of local culture and beliefs. Political ecology's concern with multiple constructions of the environment is relevant here, but we also need to focus on differences among individuals within groups in relation to perceptions and beliefs about pesticides, not just among groups defined by such characteristics as class, age, or gender. How, then, do we account for significant variability in pesticide use at the local level?

Pesticide misuse must also be understood in relation to two characteristics of all farming communities—"individuality" and "experimentation." Johnson (1972) employed these concepts to illuminate sources of innovation and change in traditional agriculture, but his analysis is also pertinent to understanding contemporary agrochemical use. Johnson asserted that the long-standing stereotype of uniformity in traditional agricultural practices was inaccurate. In contrast, he declared, considerable diversity is evident in farming practices among community members, reflecting patterns generated by "individuality" and "experimentation." Individuality is the tendency for villagers to make agricultural decisions based on their own preferences, experiences, needs, and perceptions, all of which vary from farmer to farmer. "Experimentation" is the tendency for farmers to experiment regularly with new cultivation techniques and crops, with the goal of improving production potential and reliability of output. These two characteristics are fundamental to farming communities in general, which function under a wide range of political-economic contexts. I would add that they are, more specifically, a reflection of the environmental rootedness of farming, which creates conditions of agricultural production that are inherently variable in space and time.

Individuality and experimentation also help illuminate the extent and variability of pesticide use and misuse within a community. Indeed, it is possible to argue that the failure of deskilling in contract farming and the persistence of individuality and experimentation in agriculture generally are critical dimensions of the pesticide misuse problem.

Organization of the Book

The first chapter traces the political-economic forces that have shaped the one-hundred-year linkage of the Caribbean banana industry to the United Kingdom market. Although many studies associate the growth of contract farming with flexible transnational corporate strategy over the last twenty-five years, in this chapter I reveal that the roots of the institution extend much further back in time. Two themes are stressed. One is the critical role of the state—in particular, British imperial policy directed at thwarting the power of United States–based transnational corporations involved in banana production in Latin America. The other is the significance of globalization in the form of the EU, its SEM, the WTO, and associated trends toward trade liberalization, all of which currently threaten the Windwards banana industry.

The environmental and political-economic forces shaping the contemporary Windward Islands banana industry and its system of contract farming are the subject of the second chapter. It begins with a discussion of the extraordinary number of environmental disasters (hurricanes, drought, pest infestations, and a volcanic eruption) that have repeatedly buffeted the Windward Islands banana industry; the significance of such crises has been ignored in the literature. I next explore the two critical elements in the commodity chain that attempt to regulate contract-farming enterprises directly—capital (Geest Industries, the large British transnational corporation that had been the exclusive marketer of Windwards bananas from 1952–1995) and the state (the St. Vincent Banana Growers' Association [SVBGA], the statutory corporation that monitors and coordinates the production of thousands of farmers on the island). Each is analyzed in relation to the central features of contract farming—penetration of the peasant production process, provision of a guaranteed market, grading standards and fruit quality determination, and the fairness of contracts. Especially important is the continually weak financial condition of the banana growers' association, which resulted from both the numerous environmental disasters experienced and Geest's exploitation of the industry, factors that prevented the Association from effectively regulating the production patterns of peasants. The analysis also reveals the key role played by substantial amounts of British aid to the banana industry, which helped serve the interests of the British state and Geest while guaranteeing the survival of the peasantry in the face of repeated disasters.

Chapter 3 begins with an examination of contemporary St. Vincent, revealing the overwhelming importance of the banana industry for the island's open economy. It also provides a historical perspective on the evolution of land-use patterns and the peasantry, revealing important continuities with the past—unequal control over land, relegation of peasant agriculture to marginal environments, and a long-standing tradition of food production and marketing, all of which affect the political ecology of banana contract farming today.

The next chapter shifts to the community of Restin Hill, which provides the setting for an exploration of patterns at the village level. This Vincentian community is highly representative of banana-producing villages elsewhere on St. Vincent and the other Windward Islands. Examination of the nature of households and the significance of gender, forms of land tenure, and unequal control over land indicates that banana production predominates in all types of households and land tenure systems.

The labor question is the subject of Chapter 5. I argue that two concepts used in the literature to characterize contract farmers—"disguised wage laborers" and "deskilled workers"—are inadequate to capture the complexity of the labor process. Both suffer from an unwarranted emphasis on the role of structural forces and control by capital and the state and inadequate attention to the critical role of human agency. I contend that such thorough control in the case of the banana industry is illusory, and a close examination of the daily activities of farmers reveals a wide range of production practices and labor recruitment strategies. Moreover, my examination of technological change, which has been ignored in the literature on contract farming, indicates patterns opposite of those associated with "deskilled workers." Understanding the labor process requires consideration of the environmental and biological basis of farming, which creates a unique labor process in agriculture that differs markedly from that in industry.

In Chapter 6 I explore the contentious issue of the impact of export contract farming on two problems widespread in developing countries—declining local food production and burgeoning food imports. A detailed analysis of the interactions among local food production, constraints in food marketing, banana cultivation, and the environmental context of agriculture indicates that conflicts in the productive sphere between food crops and bananas are not to blame; indeed, certain environmental characteristics of banana production actually facilitate food production. Rather, the contrasting political-economic contexts of food production and banana contract production are responsible for the decline of local food crop cultivation and growing food import dependency.

In Chapter 7, which highlights the environmental question, I first discuss the political-economic forces contributing to growing pesticide use—in particular, the roles of capital, the state, and the market. Although the literature emphasizes pesticide misuse, I reveal a wide range of beliefs and practices associated with the application of these agrochemicals—patterns that represent the failure of capital and the state to control thoroughly the production activities of peasants and instill uniformity in the labor process. The wide range in pesticide-related beliefs and practices must be understood in relation to the highly variable nature of the environment and its effects on agricultural strategies.

In the Conclusion I argue that it is unproductive to apply models developed to understand industrial restructuring to the realm of agriculture. Also, the nature of contract farming and its impacts on the peasantry are more varied than the literature indicates. Although a political-ecological perspective helps illuminate these issue, I assert that further modifications of this perspective are necessary. One change required is greater attention to variations in resource-use patterns among individuals, which has important implications for understanding human-environment relations. Another is placement of even more emphasis on the creative significance of the environment itself. Although political-economic forces are certainly crucial in influencing agriculture in general and contract farming in particular, we also need to emphasize the environmental rootedness of agriculture, which makes control over the labor process more difficult in agriculture than in industry and creates a wide range of behavioral responses that thwart efforts to instill uniformity in the production process. Thus, the unique nature of the farming experience has crucial theoretical implications.

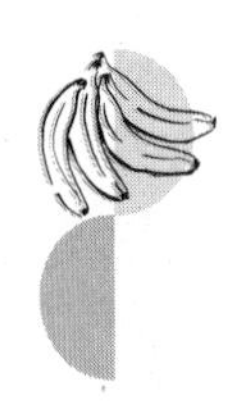

One

The History and Contemporary Context of the Windward Islands Banana Industry

The development literature is inundated with phrases such as "globalization," "restructuring," "the new international division of labor in agriculture," and other concepts intended to convey fundamental, contemporary changes in the post–World War II global economy, with especial reference to the period of perpetual crisis and upheaval since the 1970s. A recent tendency in the literature on contract farming is to link the growth of this institutional form to these broader processes (see Sanderson 1986b; Bonanno 1991; McMichael 1994c). For example, Watts (1994b: 248) asserts that "[c]ontract production in Africa as elsewhere is one manifestation of the late twentieth-century restructuring of agriculture that can only be fully comprehended as a global phenomenon." Thus, post-Fordist trends toward corporate flexibility described in the literature on industrial restructuring—growth in the importance of niche markets, competition based increasingly on considerations of quality, product differentiation, and vertical disintegration and the concomitant expansion of subcontracting networks—are also viewed as propelling the rise of contract farming (Watts 1994a,b).

But a focus exclusively on the contemporary era hinders our understanding of the complex nature of contract farming in general and the Windward Islands banana industry in particular. Indeed, Watts (1994b: 256) is careful to point out that current processes associated with restructuring of the global economy do not explain the origins of contract farming. Thus, to place the industry in the proper perspective, our analysis must extend further back in time to trace its historical developments.

At the same time, the Windwards banana industry is subject to the winds of contemporary change, particularly in relation to forces associated with globalization. Especially relevant to the Windward Islands banana industry is the growth of regional economic organizations—in this case the European Community (EC) and its successor, the European Union (EU) and its Single European Market (SEM)—which emphasizes trade liberalization within its borders.[1] The complex and changing patterns linking the British state and the Windward Islands to supranational institutions have had a fundamental influence on the history and viability of the Windwards banana industry.

Origins of the Windward Islands Banana Industry

The story of the Windward Islands banana industry and its relation to the U.K. market (the contemporary destination for Windwards bananas) begins neither in the Windward Islands nor in the United Kingdom. Rather, it starts elsewhere in the Caribbean—Jamaica—where a nascent banana export industry that was focused on supplying the United States market developed in the 1860s and 1870s. One of the enterprises involved in the early trade was the Boston Fruit Company, which grew rapidly and established plantations on the island. It subsequently merged with the banana-producing interests of Minor Keith, who had extensive plantations in Central America, to form the United Fruit Company in 1899 (later known as United Brands and today as Chiquita Brands International) (Sealy and Hart 1984). By the turn of the century, the banana industry, now dominated by United Fruit, was a significant component of the Jamaican economy, contributing 35 percent of exports (Beaver 1976: 23).

The British government, however, was growing increasingly concerned about U.S. corporate influence in the region. Thus in 1901 it provided a subsidy to the large British shipping firm of Elder Dempster for operation of a refrigerated service linking Jamaica and the United Kingdom to bring the fruit back from the colony.[2] With the arrival of regular shipments of bananas from Jamaica, it appeared that American control and domination of the Jamaican industry had finally been broken. But such independence was short-lived. Elders and Fyffes, the British firm created to take over the trade from Elder Dempster, soon ran into financial difficulties, enabling United Fruit to gain increasing control over the firm and to acquire all its share capital by 1913 (Beaver 1976: 51–52). The ironic result was that a trade established to counter the United States' influence in the region was now under the control of a U.S. corporation. United Fruit's subsidiary (eventually referred to simply as Fyffes) was soon supplying the British market with bananas not only from Jamaica but also from Central and South America.

As Trouillot (1988) points out, the British government was alarmed by this state of affairs. The Imperial Economic Committee's (1926) influential report on fruit production and trade in the British Empire highlighted this concern, lamenting the fact that "[a]n organization under American control monopolises the whole supply of bananas from Central America and Jamaica to the United Kingdom, thereby controlling the sales of 23 out of the 30 bananas consumed per head of the population in 1924" (13). Specifically, the Imperial Committee complained that such a virtual monopoly was not in the best interest of the British consumer, the British government, or Jamaican farmers. Of particular concern was the fact that the purchase of such fruit required the expenditure of U.S. dollars, which contributed to Britain's balance of payments problems.

The Imperial Committee also focused on the increasing complexity of the fruit trade, urging that banana growers in the British Empire be organized to facilitate production and marketing:

> There is no organized method of expressing the views of banana growers in any particular Colony, and in the establishment of a shipping service it would be most important for the shipping line to be able to negotiate with some representative body which was able to guarantee regularity and adequacy of supplies. There is, moreover, the educational value of producers' organisations in furthering the adoption of improved agricultural methods. [264–65]

It furthermore advised that such organizations "should in some way be under Government auspices to ensure that they were thoroughly representative in character and that their funds were applied to the purposes for which they were contributed" (265). At the same time, the Imperial Economic Committee also appealed for better quality standards in the grading, packing, and presentation of fruits. It concluded that financial assistance should be provided to help establish associations of banana producers in the British colonies (36).

Clearly, the major goal of the Imperial Committee was to lessen the grip of United Fruit on the British banana market. Its suggestion of forming banana growers' associations to negotiate with shippers was related to its additional recommendation to encourage a British shipping firm to enter the trade and begin importing bananas not only from Jamaica but also from the Windward Islands, especially St. Lucia and Grenada, to help supply the U.K. market.

But of particular interest is that the conceptual seeds of the present-day Windwards banana industry based on contract farming were already firmly planted in the report over seventy years ago—long before the institution started to spread in developing countries. Key recommendations in the report—a statutory corporation that would regulate the industry, negotiate with shipping interests, and encourage improvements in fruit quality and packing; farmer education and training; and the provision of financial aid—all characterize the industry today. Clearly, contract farming in the Windwards banana industry did not arise on the scene in the post–World War II era without significant roots in the past.

Although the Imperial Economic Committee was concerned about the supply of bananas to the British market and hoped that the Windwards would help alleviate the problem, the first sustained Windwards-wide banana export industry was directed at the Canadian market.[3] Starting in 1933, the Canadian Banana Company, under the control of none other than United Fruit, began shipping bananas to Canada, the trade encouraged by reciprocal agreements between Canada and the Windward Islands (Spinelli 1973; Momsen 1992). As the Canadian Banana Company refused to negotiate with individual growers, statutory corporations were established on each island to promote development of the industry and negotiate with the shipping company, a pattern similar to that recommended by the Imperial Economic Committee (Mourillon 1978: 9). These banana growers' associations each initially signed five-year contracts with the company, which agreed to buy all bananas considered suitable for export.

In the case of St. Vincent, the St. Vincent Banana Association was formed in 1934 and empowered as the exclusive exporter of bananas from the island. Both large-scale and peasant growers participated in the industry, with peasants making up the majority of growers. Compared with the industry today, banana production was not extensive, covering at most eleven hundred acres (Spinelli 1973: 184). Although farmers had a guaranteed market, the system cannot be considered as an example of true contract farming because no evidence is available to indicate that capital or the state intervened in the production process.

By the late 1930s, initial optimism concerning the trade's prospects began to fade. Transportation problems within the Windwards, owing to the rudimentary road systems, contributed to the bruising of fruit, and as supplies increased, the company became increasingly selective in determining what was suitable for export (Mourillon 1978: 13). Another major problem was the variety chosen for export: the "Gros Michel" banana (*Musa* AAA Group) was highly susceptible to Panama disease, a fungal infestation. Just as it was devastating Gros Michel plantings in Jamaica and elsewhere in the Caribbean and Latin America, Panama disease took its severe toll on the Windwards industry. The declining industry finally collapsed in 1942, as disruption of shipping because of naval warfare severed the link to the Canadian market.

While the Windwards struggled to develop their own banana export industry, important changes occurred in the U.K. market. Specifically, the British government established the first in a long series of import policies that provided preferential treatment for banana producers within the empire—policies that would subsequently be essential for the growth and survival of the contemporary Windward Islands banana industry. In 1932, it placed a tariff of £2.50 on each ton of bananas imported from non-empire sources, giving a considerable competitive edge to empire growers and enabling Jamaica increasingly to dominate the U.K. market. Before 1929, Jamaica supplied only 27 percent of that market, but by 1935, benefiting from the preferential tariff, it controlled 78 percent of that trade, increasing its dominance to 83 percent by 1938 (Sealy and Hart 1984: 87). But in 1940, with the onset of World War II, the government in the United Kingdom terminated all imports of bananas, curtailing the expansion of the Jamaican industry.

The banana export industries in both Jamaica and the Windwards did not revive until after World War II. With the conclusion of the war, the British Ministry of Food now became the exclusive importer of bananas, an arrangement that continued until 1953. The first bananas to arrive in the United Kingdom were from Jamaica in December 1945 (Beaver 1976: 83). But Jamaica was not able to supply all the needs of the now controlled British market, as its industry had declined during the war years. Thus, by 1948, Jamaican banana exports to the United Kingdom had fallen to less than one-third their prewar average (*West India Committee Circular* 1948: 187). This undersupplied British market provided the context for the emergence of the contemporary Windward Islands banana industry.

At the same time, the British made a decision that did not appear momentous at the time but would later have significant implications for the labor process and technological change in the Windwards. Specifically, in 1948 it designated the "Lacatan" variety of banana of the "Cavendish" group (*Musa* AAA Group) as also suitable for importation (*West India Committee Circular* 1948: 187). In contrast to the Gros Michel, which was the mainstay of the British market, the Lacatan was resistant to Panama disease. But the Lacatan had weaknesses of its own, one of which was susceptibility to another fungal infestation, leaf spot disease, or yellow Sigatoka leaf spot, but this was a less serious threat to bananas than Panama disease. A particularly crucial implication is related to the characteristics of the fruit of the Lacatan. Not only is the peel more tender than that of the Gros Michel, but the fingers of the hands of bananas do not lie as compactly on the stem (Hart 1954: 228). To those uninitiated into the intricacies of the banana trade, such differences would appear minor, hardly worth mentioning. In reality, both drawbacks make the fruit more susceptible to bruising during handling and transportation, a serious problem because the market became increasingly selective over time in relation to fruit quality.[4] Coping with this susceptibility to bruising has led the Windwards into an endless series of technological innovations that have made the production process much more complex over time, a pattern that is the opposite of what one would expect from the concept of deskilling.

The year 1948 also marked the revival of the Windwards banana industry. Paddy Foley from Dublin and a Liverpool fruit merchant, Geoffrey Band, began negotiating to purchase all exportable bananas of the Lacatan variety from the Dominica Banana Association for fifteen years (see Mourillon 1978: 15; Davies 1990: 184). They started shipping bananas in 1949, purchased land to grow the crop to supplement their supplies, and formed Antilles Products to handle the business. Initial shipments went to Dublin and the European continent but soon entered the undersupplied United Kingdom market as well (*West India Committee Circular* 1950: 110). Antilles subsequently contracted with St. Lucian growers in 1951.

The question remained whether the British market would provide the demand needed to spur the growth of the nascent Windwards industry. Indeed, many children had grown up in the United Kingdom during the war without ever seeing a banana. A minor incident in 1950 helped quell any doubts about potential demand in Britain: "[O]n December 6th a ship arrived at Liverpool from Sierra Leone with a cargo of bananas of which 78,000 were too ripe to be taken away. About 3,000 dock workers were told to eat all they could, and disposed of 37,000, an average of over 12 bananas each, before arrangements were made for disposing of the remainder" (*West India Committee Circular* 1951: 20). The English had not forgotten their love of bananas.

At the end of 1952, the British government announced its intention to cease direct purchases of all imported bananas and return the trade to the private sector and to remove all price and distribution controls on the fruit.[5] It also announced an "open general license" that permitted the importation of bananas from sterling areas without the need to apply for special import licenses. In contrast, importation of so-called dollar bananas, bananas produced in Central and South America (except Brazil) that required the expenditure of U.S. dollars and thus were a drain on foreign currency reserves, required the issuance of specific licenses.[6]

In the same year, another crucial change occurred in the Windwards (Trouillot 1988: 164). Antilles was having trouble arranging shipping to meet its commitments to purchase bananas and thus sold its interests to Geest Industries, Ltd., a rapidly growing British firm which had roots in Holland and which was involved in the vegetable and horticultural trade.[7] Geest was receiving requests for bananas from its customers in the United Kingdom and was interested in obtaining its own supplies, as Fyffes monopolized the market at the time. The British government also encouraged Geest to enter the trade (Roger Hilborne, pers. comm., 23 January 1991). Antilles' weaknesses provided a ripe

FIGURE 1.1. Windwards Banana Exports, 1954–1996

Sources: Grossman 1994; WINBAN annual reports; WIBDECO 1997.

opportunity, especially in light of the British government's 1952 decision to return the trade to the private sector (Davies 1990). In 1954 Geest signed ten-year contracts with all four banana growers' associations in the Windwards, guaranteeing to purchase all bananas of exportable quality and setting the minimum acceptable weight for a bunch at eighteen pounds. This guaranteed market, along with initially high prices for bananas, spurred rapid growth in the Windwards industry (see Figure 1.1), bringing a moderate rise in prosperity in the Windwards in the second half of the 1950s and leading to the popularity of the phrase "green gold."

The structure of the contemporary Windwards banana industry was now firmly in place. The banana industry on each island was regulated by a banana growers' association that had the exclusive right to export the fruit. Each island's association signed contracts with Geest, which, in turn, had exclusive rights to purchase all bananas from the Windwards. Moreover, the British market was now the clear destination for Windwards fruit.

The British Market and the Changing Regulatory Framework in the Postwar Era

Except for brief flickers of prosperity in the 1950s and 1980s, the Windward Islands banana industry in the postwar era has been in a precarious situation, limping from crisis to crisis. Recurrent drought and windstorms and occasional pest infestations from leaf spot, nematodes, and insects have plagued the industry and impeded growth.[8] But the real potential threat has always been from Latin American growers, who have been able to produce better-quality bananas at significantly lower cost than have growers in the Windwards. They benefit from much more advantageous environmental conditions, one dimension of their competitive edge. Not only are they outside the hurricane belt, but they also have more favorable terrain—large stretches of flat, fertile land over which it is easier to transport bananas without bruising them, a substantial difference compared with the rugged terrain in the Windwards. Major economic advantages exist as well, including vertical integration, economies of scale from large-scale production, and more widespread use of irrigation and more intensive application of agrochemicals (Ellis 1975: 47).[9] Moreover, wage rates are considerably lower in Latin American countries, reflecting greater poverty, lower living conditions, even more marked extremes in the control over wealth, and a history of political repression that has contributed to the immiserization of the wage labor force (Burbach and Flynn 1980). Ellis (1975: 47) highlighted the comparative differences between Latin American producers and those in the Windwards as they existed in 1971: yields per acre in Latin America were four to six times higher, whereas Windwards' labor costs were 50 percent higher (see also Beckford 1967: 2). More recent comparisons indicate that substantive economic differences remain. One report notes that yields in the Windwards average six tons per acre, whereas in most of Latin America they are sixteen tons (WINBAN 1990: 12); a different source indicates that yields in Central America are over twenty-six tons per acre (Matthew 1992: 7). Another study reveals that wage rates are three times higher in the Windwards compared with Colombia, a major dollar banana producer (WINBAN 1987).

The survival of the Windwards banana industry in the face of an ever changing series of economic and environmental pitfalls and the competitive threat from dollar producers has not been a matter of luck. Certainly, much of the success of the industry reflects the determination and resilience of peasant producers themselves, who make up the majority of growers. But the key role has to be assigned to the British regulatory framework governing the importation of bananas and to continual infusions of British foreign aid provided to the Windwards industry.

The initial rationale underlying British support for the Windwards banana industry was concern about its weak balance of payments situation, a result of the accumulation of war debts to the United States (Marie 1979; Welch 1994); facilitating imports from sterling areas such as the Windwards would help preserve scarce currency reserves. Indeed, the move to encourage imports from the British Empire was not limited to bananas but extended to a wide range of agricultural commodities at the time (see Cowen 1986).

Furthermore, supplies from Jamaica, Britain's main prewar supplier, still had not fully recovered. By 1954, Jamaican banana exports were only 60 percent of prewar levels (Commonwealth Economic Committee 1954: 90), and thus the market remained undersupplied. In addition, the British viewed the banana industry as a possible solution to alleviating extensive poverty among peasants in the Windwards and contributing to social stability (Thomson 1987) and as a means of facilitating economic diversification. The political issue of lessening dependence on United Fruit–controlled Fyffes was also significant (Trouillot 1988).

A key aspect of the protective umbrella meant to assist the growth of the Windwards banana industry had already been in place since 1932—the imposition of the tariff on nonsterling bananas of £2.50 per long ton, but the effectiveness of that tariff had eroded over time with inflation. Thus, to strengthen protection for sterling banana producers, the British government raised the tariff in 1956 to £7.50 per long ton, essentially restoring the 10 percent ad valorem preference established initially in 1932 (*West India Committee Circular* 1956b: 89). In contrast, sterling producers, such as the Windwards and Jamaica, could ship unlimited amounts of bananas into the United Kingdom duty free. Yet the main potential threat to the Windwards, Latin American dollar producers, would not be effectively hindered by the preferential tariff of £7.50 per ton, given their very low costs of production. Consequently, the British government also imposed an additional barrier—a quota limiting the amount of dollar fruit that could be imported to only four thousand tons per year, an amount that could be raised when production from sterling areas was insufficient to supply the needs of the British market. In fact, the first importation of dollar fruit in the postwar era was not permitted into the United Kingdom until 1959.

The protected market, coupled with comparatively high banana prices and the guaranteed market, encouraged the Windwards to expand exports rapidly (Figure 1.1). By 1959, banana imports into the United Kingdom totaled 337,000 tons, surpassing for the first time the prewar high of 327,000 tons achieved in 1937 (*Chronicle of the West India Committee* 1960: 130). But production was also increasing from Jamaica, and by 1956, Geest was already warning the Windwards about increasing competition in the British market and the need for better-quality fruit, a refrain that would be replayed over and over again in subsequent years (*West India Committee Circular* 1956a: 101).

To encourage the growth of the industry and to deal more effectively with Geest, the four Windwards banana growers' associations decided a regional alliance was essential. Thus, in 1958, they formed the Windward Islands Banana Growers' Association, commonly known as WINBAN, which became a central element in the industry. Headquartered in St. Lucia, it assumed the functions of negotiating with Geest, monitoring the British banana market, administering various aid programs and insurance schemes, conducting research, and occasionally purchasing inputs in bulk for the associations. But optimism for the industry's future surrounding the establishment of WINBAN soon began to fade.

The growth rate of the Windwards banana industry tapered off in the early 1960s, as increased supplies from both Jamaica and the Windwards resulted in lower banana prices in the U.K. market. A period of malaise seemed to be setting in among farmers, but the fortunes of the Windwards then appeared to rise significantly in 1963. Another sterling banana producer, the Cameroons, which supplied approximately 20 percent of the British market in 1962 (*Chronicle of the West India Committee* 1963: 432), decided to leave the British Commonwealth. With termination of its preferential access to the British market, its banana shipments dwindled as a result of the £7.50 tariff it now faced.

But the growth potential created by the loss of Cameroons production was limited. Continued expansion in supplies from both the Windwards and Jamaica eventually caused a glut on the British market, culminating in the so-called banana war of 1964–66 between the two major suppliers. Fyffes and Geest both tried to unload as many bananas as possible on the U.K. market in an intense battle for market share, thus causing prices to drop precipitously to a postwar low (Beckford 1966).

The crisis was finally resolved in 1966, when the Windwards and Jamaica negotiated a market-sharing agreement, an arrangement possible only because of British consent. According to the 1966 agreement, the Windwards would limit their exports to 48 percent of market requirements, while Jamaica would seek no more than 52 percent (Beckford 1967: 5).[10] Most important, the effectiveness of such voluntary action was predicated on the two marketing agents, Geest and Fyffes, not seeking supplies elsewhere. Thus, the respective parties incorporated clauses into their contracts whereby they adopted the "principle of exclusivity," agreeing not to purchase bananas from other sources unless their respective suppliers were unable to meet their market requirements.

The banana war also heightened awareness of competition, and a key factor in competitive success in the banana market has always been fruit quality, which is determined in large part by the physical appearance of bananas. Thus, variations in quality affect prices received for bananas. Unfortunately for the Windwards, the quality of the fruit they exported to the United Kingdom declined considerably in the 1960s, as indicated by the percentage of bananas graded by Geest as "specials" and "bests," the two highest-quality rankings at the time. In 1961, that percentage, according to Geest evaluations in the United Kingdom, had been over 70, but by 1968 it had fallen to a low of only 35 (SVBGA 1968: 2). The decline in fruit quality put further downward pressure on the prices growers received. Moreover, the British government threatened to increase the quota on the amount of Latin American fruit allowed into the country if the Windwards did not improve the quality of their exports—a threat which was to be repeated many times in the future and which contributed to the drive toward technological change in production at the farm level.

New marketing problems were also on the horizon. In 1969, Fyffes, expressing dissatisfaction with declining output and the low quality of Jamaican fruit, unilaterally abrogated its contract with Jamaica. The implications were negative not only for Jamaica but also for the Windwards. The market-sharing agreement of 1966, which had brought stability to the market, was predicated upon the adherence of both Geest and Fyffes to the "principle of exclusivity," which Fyffes was now unilaterally rejecting. After terminating its contract with Jamaica, Fyffes began importing lower-cost bananas from the Ivory Coast and Suriname into the United Kingdom, which once again put downward pressure on prices from 1970 to 1973. Such non-Commonwealth sources had once been hindered somewhat by the 1956 increase in tariffs on such fruit to £7.50 per ton, but the influence of that preference had declined considerably by 1970 as a result of inflation and devaluation of the British pound in 1967.

Events in Jamaica were not the only concern for the Windwards industry, which was to confront its most serious set of crises in the first half of the 1970s. Export production trailed downward from a peak of 201,903 tons in 1969 (which constituted 53 percent of the U.K. market) to 91,836 tons in 1975 (only a 30 percent share) (Ellis 1975; Persaud 1981). A variety of environmental shocks contributed to the downturn. Recurrent, severe drought and windstorms reduced profits and production in the Windwards generally, while the banana borer, an insect pest that had become resistant to the pesticides then being used, further crippled output in St. Vincent. Moreover, real banana prices (prices adjusted for inflation) for farmers fell in the first part of the 1970s, as data for St. Vincent indicate (Felton 1981a) (see Figure 1.2). At the same time, costs of labor and imported agrochemical inputs were rising. To make matters worse, the Arab oil embargo in 1973–74 sent the price of imported fertilizer, the main chemical input in banana production, skyrocketing from EC$230 per ton in 1973

FIGURE 1.2. Banana Prices for Farmers, Adjusted for Inflation, 1964–1994

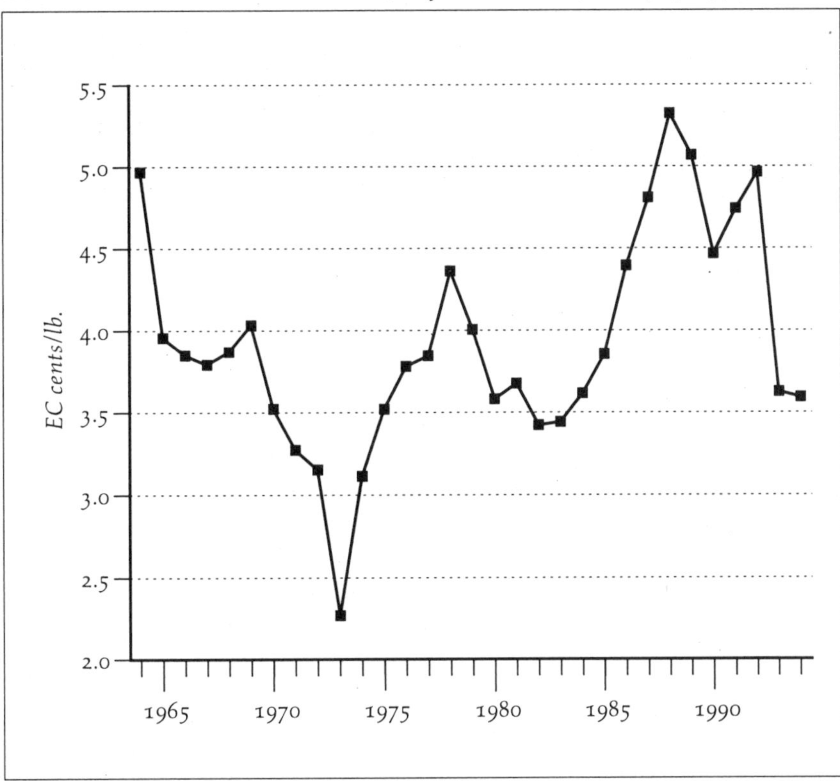

Sources: Felton 1981a; Grossman 1994; SVBGA annual reports; SVG Statistical Unit digests of statistics and files.

to EC$560 per ton in 1974 (SVBGA files).[11] Higher oil prices also resulted in increased shipping costs. As the price of agrochemicals rose and farmers' incomes fell, growers reduced their use of fertilizers and pesticides, leading to further production declines. Many farmers, disillusioned with the increasingly poor returns, stopped commercial banana production altogether and either switched to alternative crops, sought off-farm employment, or emigrated.

The swirl of change was not about to abate with Britain's successful application to join the EC in 1973.[12] One of the major goals of this regional organization, originally formed by six European countries in 1957 under the Treaty of Rome, was to reduce trade barriers among members. The implication of membership in such regional economic entities is a decline in the ability of nation-states to formulate their own economic policies, a basic dimension of globalization. In the case of bananas, the import policies of member states were subject to the rules of the EC, and different policies had been approved for different members.[13] Thus, West Germany was allowed to import dollar bananas duty free, whereas imports of such fruit into the other member states were subject to an EC-wide ad valorem external tariff of 20 percent. In contrast, France had a preferential system that reserved two-thirds of its market for its overseas territories (overseas *départements*) in the Caribbean, Martinique and Guadeloupe, and the remainder for its former colonies. Britain's application to join the EC caused considerable consternation in the Windwards because banana producers were uncertain how Britain's membership in the organization would affect its preferential system of tariffs and quotas (Spector 1967; WINBAN 1967; Sutton 1984), further reducing farmers' incentives to produce bananas in the early 1970s.

The new regulations imposed by the EC on the United Kingdom diluted the traditional protection it provided to its sterling producers, but the worst fears of the Windwards—loss of the quota on dollar fruit—were finally allayed. The United Kingdom was allowed to retain its quota and impose the EC-approved 20 percent ad valorem tariff on such fruit. But Britain's preferential tariff on non-Commonwealth fruit from both overseas territories of EC member states, such as the French Caribbean islands of Martinique and Guadeloupe, and "ACP" countries—former colonies of EC members in Africa, the Caribbean, and the Pacific—had to be gradually eliminated, which presented a problem for the Windwards. Several ACP countries could produce bananas at a lower cost than the Windwards; without any protection from such sources, the Windwards feared that prices might drop and that they might lose market share. Furthermore, the nearby French islands of Martinique and Guadeloupe enjoyed a highly protected market in France, enabling them to dump surplus production elsewhere at lower cost, another potential problem.

Certainly, the Windwards industry was on the brink of collapse in the early 1970s, and future prospects seemed dim. The mood of the islands was captured succinctly by the premier of St. Lucia, John Compton (1974: 9), who lamented:

> Over the past nine years we have fought together many a battle to ensure the survival and advancement of the Banana Industry upon which the economic and social stability of these Windward Islands so heavily depends. During these years, the Industry has undergone many changes of a fundamental nature and the Industry which was once glorified as "green gold" has seen the "gold" disappear from the "green."

But the fortunes of the Windwards once again changed. As it had done in the past, British policy again came to the rescue. The solution had its seeds in the previous Fyffes-Jamaica dispute. The British government had appointed Lord Denning to mediate the conflict in 1970, and one of his suggestions was to create a "managed" banana market in the United Kingdom in which imports were controlled to ensure that the Windwards and Jamaica were the main beneficiaries—similar to the protective regime France had instituted to protect its own overseas territories and colonies. The British government rejected the idea at the time, but the turmoil in the banana market in 1972 and 1973 led it to reconsider its position. Thus, in 1973 it established the Banana Advisory Committee (later called the Banana Trade Advisory Committee), which was chaired by the Ministry of Agriculture, Fisheries, and Food (MAFF) and included the major marketers of Jamaican (Fyffes and Jamaica Producers) and Windwards (Geest) bananas and representatives of growers from both Jamaica and the Windward Islands. The official purpose of the committee was to advise the British government concerning likely supplies from the Windwards and Jamaica and other ACP producers and on whether shortfalls from these sources would necessitate increasing the amount of dollar fruit imported beyond the quota of four thousand tons. But a much more important result of the establishment of the committee was an informal agreement among Geest, Fyffes, and Jamaica Producers not to import more than the market could bear and to divide the British market among them.[14]

The committee, however, did not set banana prices, and competition among the three importers existed in relation to supplying the retail and wholesale trades. Nonetheless, it had a significant, positive influence on prices by coordinating imports to prevent oversupplying the market. According to Harry Atkinson (pers. comm., 21 June 1989), formerly president of WINBAN and chairman of the St. Lucia Banana Growers' Association, the actions of the committee, without doubt, saved the Windwards industry from collapse, with the informal agreement continuing to work satisfactorily for the three importers for almost the next twenty years, until the early 1990s. The committee also benefited the three importers themselves. Until the late 1980s, it regularly allocated to them the large majority of supplementary dollar licenses issued when imports from both Commonwealth and non-Commonwealth ACP countries were inadequate to supply the British market; for example, from 1975–76 to 1981–82, they obtained between 85.6 and 93.7 percent of such licenses (Matthew 1984: 37).

Further heartening news came in 1975, this time from the EC, which signed the first Lomé Convention with forty-six ACP countries, including the Windwards. This broad-ranging agreement contained provisions for aid and trade and also had a special section on bananas, known as the "Banana Protocol." This key provision for the ACP banana producers guaranteed that their banana exports to the EC would not be placed in a less favorable position with regard to their market access and market advantages than they were currently enjoying or had enjoyed in the past. This Protocol, reiterated in subsequent Lomé Con-

FIGURE 1.3. Exchange Rates, EC Dollar to British Pound, 1976–1996

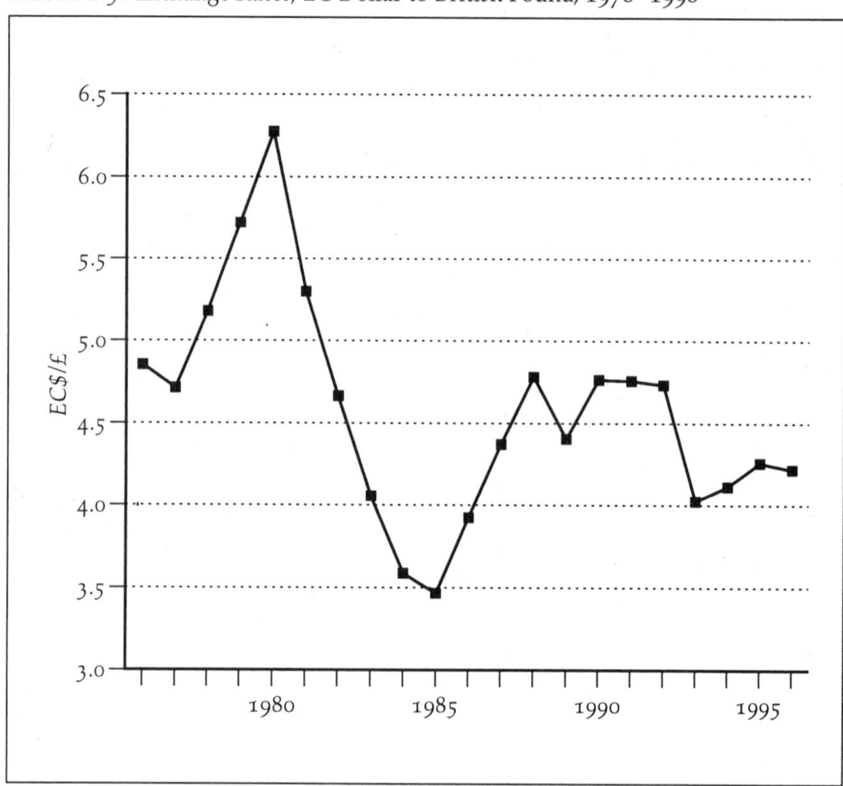

Sources: WINBAN and WIBDECO annual reports; Slater 1996.

ventions, essentially empowered Britain and the other EC countries to continue providing protection for their "traditional" banana suppliers.[15]

The Lomé Convention inspired increased optimism in the future of the Windwards industry. And production and prices also finally started to rebound in the second half of the 1970s, facilitated in part by greater stability in the market imparted by the actions of the Banana Trade Advisory Committee and the cessation of the long drought. But prolonged periods of stability or prosperity have simply not been characteristic of the history of the Windward Islands banana industry, and this period was no exception. A series of environmental disasters soon struck, again disrupting growth of output in the Windwards. In 1978 and 1979, Dominica was plagued by a severe leaf spot epidemic. In 1979, a volcanic eruption on St. Vincent ruined much of the island's crop, and Hurricane David totally devastated banana fields in Dominica and did significant damage in St. Lucia. Then, in 1980, Hurricane Allen destroyed over 90 percent of the entire Windwards' banana crop.

Economic problems soon followed. Production in the early 1980s was also constrained by inflationary pressures, as another crisis in the Middle East, this time associated with the fall of the Shah of Iran, once again led to oil price hikes in 1979 and 1980 that increased the costs of imported fertilizers and shipping. And unfavorable currency exchange rates in the first half of the 1980s further lowered returns (see Figure 1.3).

Finally, in the latter half of the 1980s and early 1990s, a period of prosperity and expanding production—the so-called banana boom—lifted the fortunes of the industry and its growers. Indeed, between 1985 and 1990, the Windwards supplied between 50 and 60 percent of the U.K. market compared with only 20 to 42 percent during the previous five years (United Kingdom Ministry of Agriculture, Fisheries, and Food files).[16] Favorable changes in the contract with Geest (discussed in Chapter 2), better exchange rates, and technological changes in harvesting and packing that improved fruit quality all contributed to higher prices for farmers.

The boom period reverberated throughout the Windwards economies. My first visit to St. Vincent in 1988 coincided with this efflorescence of activity. On the island, private housing construction increased substantially, as many farmers upgraded from the older, wooden "board houses" to the more expensive concrete "block houses." Imports of vehicles rose, intended for both farmers and those transporting fee-paying passengers in vans from villages to Kingstown, the nation's capital (Map 2). Retail activity in Kingstown blossomed; sales of consumer items and household appliances—refrigerators, gas stoves, and televisions—rose significantly, with such items becoming increasingly widespread in the countryside. Bank loan activity for both residential construction and land purchases exploded. Villagers were quite positive about the benefits of banana farming, despite its high level of labor intensity. Similar increases in prosperity were evident in other banana-producing villages elsewhere in the Windwards (see Barrow 1992). Indeed, from the vantage of this short time horizon, one could conclude that prosperity had, at last, arrived for the peasantry. Without doubt, small-scale banana producers were enjoying a level of prosperity unprecedented in the difficult history of the Vincentian peasantry. But a knowledge of the history of the banana industry specifically and of commodity production in developing countries generally imparts a more skeptical view, and history, once again, repeated itself.

In spite of growth in the British market—banana imports from all sources expanded from 303,400 metric tons in 1980 to 467,400 in 1990—production in the Windwards rose too rapidly for its share under the informal agreement among the three major importers, and consequently, the Windwards, through Geest, began marketing some bananas in Italy as well, where the fruit com-

MAP 2. St. Vincent

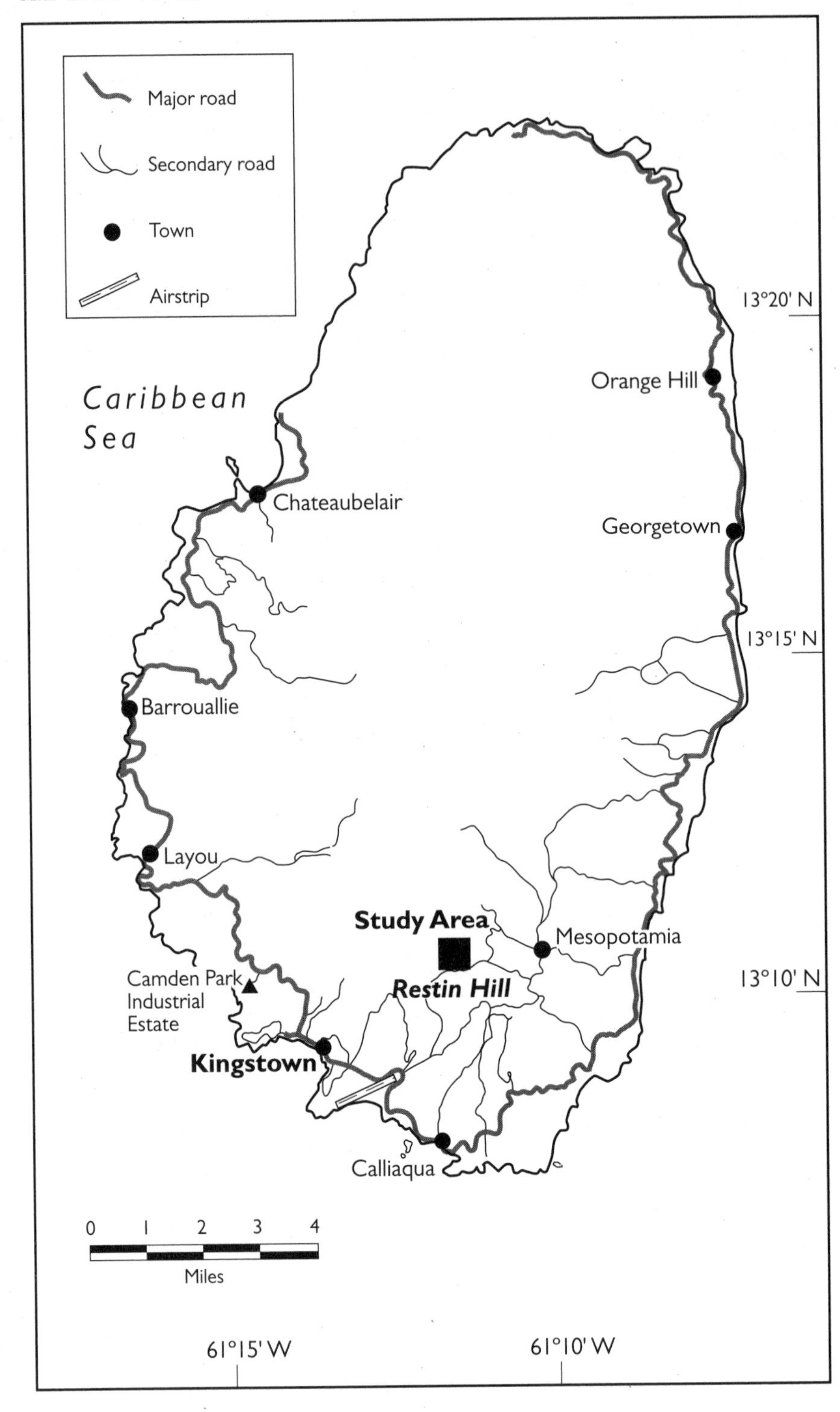

manded lower prices (WINBAN 1981a, 1991). Contributing to the problem was the decision by the British government in 1988 to raise the quota on Latin American fruit to spur increased competition in the market and to try and force the Windwards to improve fruit quality; it thus set the new minimum dollar quota at thirty thousand metric tons for 1989 and also indicated that it would increase the minimum quota in subsequent years (Ministry of Agriculture, Fisheries, and Food 1990). Had the British not changed the dollar quota, the benefits from the Windwards banana boom would have been even more impressive.

By late 1992, the traces of prosperity were evaporating rapidly. A sharp decline in exchange rates was depressing growers' returns. But as significant as the decline was, another change had even more momentous and ominous implications for the future of the Windwards industry.

Bananas and the Single European Market

Once more the process of globalization was impinging on the lives of Windwards growers—changes that are now threatening the very viability of the industry itself. The changes were not unexpected, and indeed, for several years in the late 1980s and early 1990s, the upcoming year 1992 was much discussed everywhere, from official government circles to village rum shops. That was the year during which the long-heralded SEM was supposed to be established.[17] The goal of the SEM was to remove all existing barriers to the movement of goods, services, labor, and capital among the twelve member countries of the EU (successor to the EC in 1993) and create the world's largest market, with 350 million consumers (Sutton 1995).[18] The rationale for the formation of this new supranational entity was to spur economic growth among its members and increase the international competitiveness of corporations within the EU, "a response to an increasingly hostile and competitive global economic environment" (Nurse and Sandiford 1995: 9).

The relationship of the EU and the SEM to the process of globalization requires explanation. Certainly, it represents a diminishing of the authority of member nation-states to regulate their own affairs, as policies linked to a broad range of trade-related and domestic issues are now controlled by joint determination of EU members. Trade liberalization is also promoted, though it is mainly among member states. But barriers were created in relation to the importation of certain items into the SEM, including bananas. Thus, the SEM does not represent trade liberalization at the global level. Nonetheless, most researchers consider the EU and its SEM to be an example of globalization because of its emphasis on expanding free trade within its borders and the decline in the autonomy of its member nation-states (Brecher and Costello 1994; Koc 1994; Gereffi 1996; Thurow 1996; compare Bonanno 1994; Watson 1994). Thurow (1996: 120) remarks: "Jumping from national economies to a one-world economy is a leap too big to make. As a result regional trading blocs are emerging as natural stepping-stones in an evolutionary process toward a truly global economy."

The implications for the banana trade were momentous. First, the EU became the world's largest market for bananas (Borrell and Cuthbertson 1991: 3), importing between 35 and 40 percent of all bananas traded internationally (European Commission 1995: 4). Even more important, the creation of the SEM meant the dismantling of the various preferential regimes that several member states, including the United Kingdom, had instituted to provide protection for both their traditional ACP suppliers and for their overseas territories that produced bananas. The outcome for the Windwards was the loss of the preferential system of tariffs and quotas in the U.K. market that had been so essential for the industry's survival. Instead, the SEM was going to incorporate one uniform set of regulations that would apply to all twelve member nations. Britain was still supportive of the Windwards, but it no longer had the control it once exercised over the fate of the Windwards industry. Now it was just one of the twelve voting member states in the EU that would determine the nature of future protection.

Resolution of SEM-wide banana regulations was one of the most contentious of the thousands of issues involved in creating the Union, "the subject of a prolonged and acrimonious debate between Member States" (Read 1994: 225). The discussions involved several conflicting policy objectives, including internal liberalization of trade, increased consumer benefits, support of traditional banana suppliers, and compliance with GATT rules on trade (219). The issue of how to deal with the importation of dollar fruit from Latin America was central to the debates (Read 1994; Nurse and Sandiford 1995). The interests of the various EU members were highly diverse, reflecting the nature of their preexisting import regimes. One group, consisting of the United Kingdom, Italy, France, Spain, Portugal, and Greece, had protected markets for their former colonies, now ACP producers, or for their overseas territories, or for both. These countries worried that unlimited supplies of low-cost Latin American fruit would drive prices down, making production unprofitable for their traditional, higher-cost suppliers and thus likely drive them out of business. In contrast, a second group, including Germany, Belgium, Luxembourg, Ireland, Denmark, and the Netherlands, had neither overseas territories producing the crop nor preexisting bilateral relations with ACP exporters, instead importing dollar fruit to supply their market needs without any quotas. These members, especially Germany, feared that quotas on Latin American fruit would certainly drive prices higher for their consumers.

Clearly, Latin American fruit was dominant in the EC. In 1992, dollar bananas supplied 62 percent of the EC's needs, compared with only 20 percent from overseas territories of EC members and 18 percent from ACP producers (European Commission 1995: Annex 1).[19] The primary exporters of dollar fruit are the powerful "big three" United States–based banana transnationals—Chiquita, Del Monte (now controlled by a Chilean company owned by Jordanian interests), and Dole, which together supplied 43.5 percent of EC needs in 1991 (Arthur D. Little 1995: 11). Indeed, Latin American fruit is predominant not only in Europe but also globally in the US$7 billion-per-year banana trade, supplying 75 percent of worldwide exports; against such a potentially overwhelming force stood the small Windward Islands, which shipped only 2.6 percent of the global trade in 1992 (Food and Agriculture Organization 1995: 131; 1996: 140).

The forthcoming SEM certainly did not go unnoticed by dollar producers. Latin America increased banana shipments to the EC by 46 percent from 1988 to 1992, from 1.64 million to 2.40 million metric tons, in an effort to capture market share and influence subsequent decisions about possible new quotas in the SEM (Nurse and Sandiford 1995: 112). Indeed, the EC market became flooded with bananas in 1991 and 1992 as a result of such actions, which led to lower prices throughout the EC (Read 1994). Moreover, the "big three" were also active in investing in ACP and EC banana producers, ripeners, and marketers in both joint ventures and acquisitions to gain further leverage in the new market (Arthur D. Little 1995; Nurse and Sandiford 1995).

The Windwards and other ACP producers, fearful of the possible dollar fruit onslaught, argued for continued protection, reflecting their inability to compete with dollar fruit in an open market. The results of open competition with Latin America would be particularly devastating for the Windwards, given their very high level of dependence on bananas, especially in the economies of St. Vincent, St. Lucia, and Dominica (see Table 1.1). Furthermore, as Nurse and Sandiford (1995: 70) indicate, approximately half of the working population in the Windwards are directly or indirectly employed by the banana industry. To bolster their plea to the EU, Windwards officials have warned that the likely result of loss of market protection and demise of the banana industry would be an upsurge in the production and trafficking of illegal drugs.

Not to be outdone, Latin American producers proclaimed that SEM quotas limiting imports of dollar fruit, particularly below their level in 1992 (2.4 million metric tons), would result in severe economic losses and increased unemployment in their own countries. And to top the threat of the Windwards,

TABLE 1.1. Characteristics of the Windward Islands Banana Industry, 1992

Indicator	Dominica	St. Lucia	St. Vincent	Grenada	Windwards
Population	71,495	138,151	106,499	90,961	407,106
Banana exports					
Volume (metric ton)	58,025	132,853	77,361	6,300	274,539
Value (EC$m)	82.2	184.9	101.4	7.8	376.3
Banana exports as % of domestic exports	56.6	60.1	49.3	17.2	53.5
Banana exports as % of GDP	19.1	17.5	18.3	1.7	15.1
Number of active growers	6,555	9,500	8,000	600	24,655
Banana acreage	12,000	16,500	12,000	1,200	41,700

Source: WINBAN 1993a: 6.

Latin American governments proclaimed that disruption of their banana industries would result in more illegal shipments of cocaine!

The issue was even more complicated because of conflicting EC commitments to other treaties. The most recent Lomé Convention, IV, in force from 1990 to 2000, incorporated the basic tenets of the Banana Protocol that was first put into effect in Lomé I in 1975 (now Protocol 5). The Lomé Conventions were the underpinnings of the various protective regimes that existed in the EC. In contrast, EC members also had commitments and obligations under GATT, one of the powerful forces of globalization directed at fostering trade liberalization (McMichael 1994c). Making matters even more complicated was the fact that the most recent round of GATT negotiations, the Uruguay Round, was being held while the new EC banana policies were being formulated, and one of the main proposals was the removal of barriers to trade in agricultural products.

In February 1993, after much vociferous debate and intense lobbying by all interested parties, the EU formulated its final policies (Regulation 404/93) for the new "banana regime," which would come into operation on July 1, 1993, and remain in effect until 2002. The regulations were, in fact, quite complex, but a brief overview of their most salient features is worthwhile.[20]

- Dollar fruit imports would have an initial annual tariff quota of 2.0 million metric tons and would be subject to a tariff of ECU 100 per ton.[21] A prohibitive tariff of ECU 850 per ton would be imposed on imports above this amount.

- Bananas from traditional ACP suppliers (St. Vincent, St. Lucia, Dominica, Grenada, Côte d'Ivoire, Cameroon, Suriname, Somalia, Cape Verde, Madagascar, Belize, and Jamaica) would be allowed tariff-free entry, but each country would be subject to an individual, nontransferable quota, based on its best year's exports prior to 1991. Amounts shipped above its tariff-free quota would be subject to a prohibitive tariff of ECU 750 per ton. The total amount of quotas for the ACP producers was set at 857,700 tons. For the Windwards, the quotas were St. Vincent, 82,000 tons; St. Lucia, 127,000 tons; Dominica, 71,000 tons; and Grenada, 14,000 tons.
- Fruit from producers in the EU (Crete) and in overseas territories of EU countries (Martinique, Guadeloupe, Canary Islands, and Madeira) would also be allowed tariff-free entry, with each country subject to an individual quota. Amounts shipped above the quota would be subject to the prohibitive tariff of ECU 750 per ton. The total amount of quotas for the EU producers was set at 854,000 tons.
- Licenses for importing lucrative dollar fruit would be controlled by allocating 66.5 percent of the licenses to traditional dollar importers, 30 percent to importers of ACP and EU fruit, and 3.5 percent to new importers.

The new banana regime was clearly a difficult compromise, satisfying none of the competing parties completely. Latin American producers complained that the quota of 2.0 million metric tons was 400,000 less than they shipped in 1992. In contrast, ACP producers, including the Windwards, hoped for a much lower dollar quota. Particularly disturbing was the fact that the EU could raise the quota on dollar fruit if conditions warranted. Moreover, the Windwards no longer had unlimited access to the European market, though since the implementation of the SEM, their levels of production have been significantly below the imposed ceilings.

But certain of these SEM regulations also had important benefits for the Windwards. As 30 percent of dollar licenses would accrue to importers of ACP and EU fruit, an arrangement intended to encourage importers to continue marketing such bananas, the Windwards gained significant leverage with Geest, as that company now depended on shipping Windwards fruit to gain access to profitable dollar licenses. Moreover, the Windwards could register as the official importer of their bananas themselves, thereby acquiring the profitable dollar licenses on their own, a potentially significant new source of revenue—a strategy that they have, in fact, implemented; indeed, the value of such dollar licenses to the Windwards generally in 1995 was EC$15 million, of which St.

Vincent received EC$3.86 million (SVBGA 1996: 19). Nonetheless, such benefits would not compensate for the increased competition that the Windwards would experience.

The new regulations of the SEM have also had important implications for the labor process at the local level. Previously, Windwards banana exports had to conform to standards set by the British government. Now they had to comply with new EU-wide regulations governing a multitude of characteristics involving quality, shape, weight, and size, which has required readjustments in production practices in Vincentian fields.

The passage and implementation of the new banana regime of the SEM did not end the swirl of controversy. Five Latin American countries—Nicaragua, Venezuela, Colombia, Costa Rica, and Guatemala—lodged two complaints with GATT panels challenging the regulatory frameworks dealing with the importation of bananas as being discriminatory and unfair restraints on trade. The EU settled this complex dispute with four of the five countries (except Guatemala) when it agreed to enlarge the dollar quota to 2.1 million tons in 1994 and 2.2 million in 1995 and reduce the tariff on such fruit from ECU 100 to ECU 75 per metric ton. The compromise, known as the Framework Agreement, reflected the force of GATT and has clearly worked to the detriment of Windwards and other ACP producers. Moreover, Germany, unhappy with the increase in banana prices at home as a result of the SEM, lodged complaints about the quota system with the European Court of Justice but was unsuccessful in its efforts.

Even more pressure has come from the United States, supported by four Latin American countries (Ecuador, Mexico, Guatemala, and Honduras), which complained in 1996 to the WTO—the international body that is the successor to GATT and is supposed to promote trade liberalization and settle trade disputes—that the EU banana regime system of licenses and quotas discriminates unfairly against the United States transnationals, even though no jobs in the United States are at stake![22] And no less than four-star United States Marine general John Sheehan, commander of all United States forces in the Atlantic and the Caribbean, warned in June 1996 that if the United States were successful in its complaint and the Caribbean banana industries consequently collapsed, the result would be increased illegal immigration from the region into the United States and a marked expansion of drug trafficking through the area, echoing previous warnings from the Caribbean in the early 1990s.

Since the implementation of the SEM, the fortunes of the Windward Islands banana industry have plummeted. Part of the problem stemmed from unfavorable weather in 1994 (a severe drought affected the Windwards and Tropical Storm Debbie devastated production in St. Lucia and Dominica), in 1995 (Hurricanes Luis and Marilyn destroyed production in Dominica), and again in 1997 (a drought in St. Vincent). Moreover, banana prices for growers have been very low from late 1992 to the present (mid-1997), the downturn resulting not only from oversupply conditions related to the decline in protection in the SEM but also from a variety of other conditions: price wars among supermarkets, or "multiples," in the United Kingdom in 1995 using bananas as loss leaders to entice shoppers (Nurse and Sandiford 1995: 41); unfavorable exchange rates from the end of 1992 to 1996; persistent fruit quality problems; and mismanagement in the banana growers' associations. Indeed, quality problems had been so severe that several major supermarket chains in the United Kingdom threatened to terminate purchases of Windwards fruit; to keep such customers, the Windward Islands have had to buy higher-quality bananas from Latin America for these customers, using some of their own dollar licenses.

As a result, growers have become demoralized, especially as labor requirements in banana production have increased as a result of mandated new technologies at the farm level, leading to a sharp downturn in production compared with the boom period before the SEM (see Figure 1.1). In fact, as conditions deteriorated in St. Lucia, banana farmers there went on a three-day strike in October 1993, seeking higher prices, which led to the killing of two growers by police. The shocking event spurred the governments of the Windwards into action, pressing the banana growers' associations to renegotiate their contract with Geest in an attempt to improve returns to growers. Moreover, it energized the effort to restructure the industry.

The decline in the industry sent shock waves through the Vincentian economy, as it has elsewhere in the Windwards. Unemployment and crime rates on the island have risen. Three commercial banks active in providing agricultural loans reported that delinquency rates increased from 8 percent to as high as 23–24 percent over the period from 1992 to 1994 (SVBGA 1995: 2). Several stores in Kingstown closed, and many others reduced operations. Similarly, residential construction activity declined considerably, negatively affecting not only builders and suppliers but also many truckers who depended on transporting building materials. Money spent in the ubiquitous rum shops in villages also fell. Some banana farmers left the island altogether, finding such jobs as landscaping in Mustique and as waiters on cruise ships. And in nearby St. Lucia, banana farmers became involved in another bitter strike in 1996, fomented, in part, by Chiquita's offer to begin buying St. Lucian bananas—a ploy viewed locally as being designed to disrupt and divide the Windwards banana industry.

The optimism of the late 1980s in the Windward Islands banana industry clearly has been replaced by a sense of foreboding in the 1990s.[23] The industry is resigned to gradually declining levels of protection—a realistic assessment given the pressures for trade liberalization and the lessening geopolitical significance of ACP producers with the evaporation of the Cold War. Furthermore, the EU is expanding—Sweden, Finland, and Austria joined in 1995—and the new members do not have traditional ties to ACP producers; consequently, their votes are not likely to be particularly sympathetic to these developing countries. Moreover, the EU expanded the annual tariff quota on dollar bananas from 2,200,000 to 2,553,000 metric tons to accommodate the consumption needs of these new members, a change that has had further negative impacts on banana prices throughout the EU (Sutton 1997: 32). And the EU itself has already signaled that major changes will be made to the Lomé Convention when it comes up for renegotiation. In addition to the influences of trade liberalization and the changing geopolitical landscape, the EU is also responding to what has become known as "Lomé fatigue"—frustration that the conventions have not already had a more beneficial impact on the ACP countries.

The response of the Windwards industry to the realities of the SEM, similar to that of almost everyone else in the 1980s and 1990s, has been to "restructure"—to make the banana growers' associations and the growers more "efficient" and "productive" and "to cut out waste" in the hopes of making them more competitive. One element of this restructuring has been to replace WINBAN with a new regional entity, the Windward Islands Banana Development and Exporting Company (WIBDECO), which was created in 1994 and which has become much more involved in the commercial operations of the industry than was its predecessor.

But whatever strategies the Windwards devise, they will most likely be devastated by the impact of the WTO's ruling in late April 1997 in favor of the complaint lodged by the United States and the four Latin American countries. In a four-hundred-page document, its length a reflection of the legal complexity of the case, a WTO panel determined that the EU's system of allocating licenses to banana importers violates international trade rules by unfairly discriminating against producers and marketers of Latin American fruit, including the big three United States–based transnationals involved in the trade (*Caribbean Insight* 1997). Although the WTO did not rule against the EU's system of preferential tariffs that also benefits ACP exporters, the tariffs alone will not be sufficient to protect ACP banana growers, such as the Windwards. The EU thus appealed the decision, but the Appellate Body of the WTO upheld the panel's determination a few months later in September. The EU subsequently declared that it would comply with the WTO findings, but exactly what will happen is unknown. The EU will likely be slow in reaching a final solution because it is divided internally into two opposing camps. One group, led by the Germans, wants to dismantle the existing banana import regime quickly to reduce banana prices in the SEM, whereas the other, led by France, the United Kingdom, and Spain, wants to preserve as much protection and support for ACP banana producers as possible. The EU has fifteen months—until January 1999—to restructure its banana import regime to comply with the WTO decision. Various alternative solutions have been suggested to help ACP growers, such as British consumers boycotting dollar bananas and nongovernmental organizations in the EU marketing Windwards and other ACP bananas as "fair trade" bananas—fruit produced under favorable social and environmental conditions—with the intention of paying a premium for such fruit. Nonetheless, the options for the EU appear limited, and such solutions will have only limited benefits for ACP banana exporters. Any changes made to EU import policies as a result of the WTO ruling will result in even more severe competition and lower banana prices—a context in which survival of the Windwards industry is doubtful. The future is also threatening because the EU's banana regulatory framework is scheduled to expire in 2002. At that time, changes will be made in its import rules that will most likely be unfavorable to the Windwards and other ACP banana exporters.

In any case, the ability to squeeze out more from the already demoralized, underpaid, and overworked banana peasantry has clearly reached its limits. The task is daunting, especially given the overwhelming power of the opposition in Latin America and its ally, the United States—along with the impact of the WTO and its mandate to foster trade liberalization. Furthermore, coping with such pressures will be especially difficult because the Windwards' current and future problems will be compounded by the numerous environmental and political-economic forces that have always placed the banana industry in a perpetually precarious position, a topic that I explore in the next chapter.

Two

Environment, Capital, and the State in Banana Contract Farming

Although several works have been published on the Windward Islands banana industry, none has considered it in relation to the institution of contract farming (for example, Marie 1979; Thomson 1987; Trouillot 1988; Nurse and Sandiford 1995; Welch 1996). Perhaps part of this oversight stems from the industry accruing the accoutrements of contract farming only gradually. Another reason, undoubtedly, is the absence of formal, written contracts between purchaser and grower. But written contracts are not the hallmark of contract farming (Watts 1994a). Rather, the central elements of the institution are the provision of a guaranteed market and intervention by capital and/or the state in the production process—both fundamental aspects of the contemporary Windward Islands banana industry. Such oversight will surely be remedied in future analyses, as the banana growers' associations are now adopting formal, written contracts.

The banana industry certainly has characteristics traditionally associated with the employment of contract-farming relationships. Production is labor intensive, quality considerations are paramount, and the crop is highly perishable. The scheduling of harvesting must be coordinated precisely with the timing of transatlantic shipping. Regularity of a guaranteed supply is also crucial, as banana ripening rooms in the United Kingdom must continually process and deliver fruit year-round to stores, which display the highly perishable commodities for only three or four days.

This chapter focuses on the forces shaping the banana industry and its system of contract farming. In particular, I emphasize the central influences of the environment, capital represented by Geest Industries, and the state in the form of the St. Vincent Banana Growers' Association (SVBGA) and the British government. The examination involves consideration of the classic dimensions of contract farming—profitability, the guaranteed market, the nature of contracts, quality determination, and intervention in the production process.

The Hazardous Environment

The tendency in the literature on contract farming is to focus on the nature of involvement by capital and the state in explaining the returns to contract farmers. But as the case of the Windwards reveals, the role of the environment is another essential element in this story. Buffeted repeatedly by environmental problems beyond its control, the industry has limped from crisis to crisis. The banana industry has, in the words of one WINBAN official (Matthew 1983: 21), "endured an abnormal share of crises." The intensity and regularity of such crises have helped place the industry in a perpetually weak financial condition, which in turn ultimately contributed to an inadequate and underfunded extension service, seriously reducing the ability of capital and the state to control the labor process and deskill labor. Moreover, the poor financial health of the SVBGA helps to explain, at least in part, why the labor process became more complex over time (see Chapter 5). This analysis thus supports the view that political ecology needs to focus not only on the impacts of political-economic forces on the environment but also on the significance of the interaction between the two domains (Grossman 1993).

The vulnerability of the banana plant to environmental shocks was captured by Popenoe (1941: 8), who wrote:

> Many of us who come to the tropics from the temperate zone have been trained to think of fruit trees as slow growing, hard-wooded plants with relatively great capacity for withstanding unfavorable conditions of environment as well as bad treatment generally. It takes us some time to realize how different the banana plant is from the fruit trees with which we have been familiar. It is not a tree at all: it is a gigantic herb capable of almost prodigious cell activity and therefore requiring abundant and constant supplies of water, heat, and plant nutrients, particularly nitrogen. It is sensitive to its environment in a high degree. It feels the shock of drought or of cool weather promptly and definitely, and reflects it in its growth and fruit.

The Windwards environment is hardly ideal for banana production. Drought is not an infrequent occurrence, a problem because banana plants require a continual supply of water for adequate plant growth and fruit production. Even after the cessation of drought, the effects continue to hinder production for another nine to twelve months. The Windwards' banana industry was particularly devastated by the long-lasting drought from 1970 to 1975, though this natural disaster has plagued the region at other times as well, including during the 1990s.

Even more potentially damaging are the numerous hurricanes, tropical storms, and less intense windstorms that frequent the region.[1] The tall banana plants are top heavy and, given their shallow root systems, are easily blown down by winds. Although banana fields can resume production six to nine months after destruction by heavy winds, the frequency of such disasters has had a constantly crippling impact on the financial health of the industry.

Added to these disasters are occasional pest outbreaks, such as leaf spot disease, that rise beyond their normal levels of damage to cause widespread losses. What has been especially disastrous for the Windwards is the regularity of the various plagues of drought and windstorms augmented by an occasional pest epidemic and volcanic eruption. A description of events in the late 1970s captures the potential combination of disasters that the Windwards industry has endured:

> It was soon realised that the projected target of 153,000 tonnes would not be achieved because of the severity of the Leaf Spot outbreak in Dominica which had begun to assume epidemic proportions in late 1978 resulting in the cutting back of 2,500 acres of the worst affected areas before the end of March, 1979. By then Dominica's production had fallen 29% below the level of the corresponding period for 1978, and by July, it was 42% below the 1978 figure.
>
> Dominica had just began to recover from the Leaf Spot epidemic which had cost a considerable sum to bring under control, when, on August 29th the fiercest storm of the century, Hurricane "David" unleashed its full fury on Dominica's banana plantations and destroyed them all.
>
> Damage ran into millions of dollars and massive international aid had to be mobilised to help rehabilitate and restore the Industry in that island. Dominica's final exports for 1979 ended with a shipment of 535 tonnes on the 23rd August and totaled 15,123 tonnes.
>
> This was a drop of more than 58% below the 1978 export figure and 64% less than the early projection of 43,800 tonnes for 1979.
>
> While all this was happening in Dominica, the Industry in St. Vincent had suffered a crippling blow on Good Friday April 13th (very appropriately called Black Friday) when the Soufriere Volcano literally rained hundred of thousands of tons of volcanic ash on St. Vincent. The damage to the banana crop was colossal. The fine ash fell between the fingers of the bananas and caused numerous minute abrasions which in turn resulted in unsightly discoloration of the fruit making them unacceptable and unfit for export. [WINBAN 1980a: 3–4]

Indeed, rarely do more than a few years go by without the Windwards experiencing at least one serious environmental hazard (see Table 2.1). This rather

TABLE 2.1. Environmental Disasters Affecting the Windward Islands Banana Industry

Date	Type	Islands Most Affected
1955	Hurricane Janet	Grenada
1958	Drought	St. Vincent
1960	Hurricane Abby	St. Lucia
1963	Hurricane Edith	St. Lucia, Dominica, and St. Vincent
1963	Hurricane Flora	Grenada
1963	Hurricane Helena	St. Lucia
1966	Hurricane Inez	Dominica
1966	Hurricane Judith	Grenada, St. Vincent, and St. Lucia
1967	Hurricane Beulah	St. Vincent and St. Lucia
1970	Hurricane Dorothy	Dominica
1970	Drought	All Windwards
1971	Drought	All Windwards
1971	Hurricane Chloe	Dominica
1972	Drought	All Windwards
1973	Drought	All Windwards
1974	Drought	All Windwards
1975	Drought	All Windwards
1975	Banana borer	St. Vincent
1976	Banana borer	St. Vincent
1978	Leaf spot epidemic	Dominica
1979	Leaf spot epidemic	Dominica
1979	Hurricane David	Dominica
1979	Volcanic eruption	St. Vincent
1980	Drought	St. Vincent
1980	Hurricane Allen	St. Lucia, St. Vincent, and Dominica
1986	Tropical Storm Danielle	St. Vincent
1987	Tropical Storm Emily	St. Vincent
1987	Drought	All Windwards
1989	Hurricane Hugo	Dominica
1990	Tropical Storm Arthur	Grenada
1991	Drought	All Windwards
1994	Drought	St. Vincent, St. Lucia, and Dominica
1994	Tropical Storm Debbie	St. Lucia and Dominica
1995	Hurricane Luis	Dominica
1995	Hurricane Marilyn	Dominica
1995	Tropical Storm Iris	St. Lucia, St. Vincent
1997	Drought	St. Vincent, St. Lucia

Sources: WINBAN and SVBGA annual reports.

TABLE 2.2. Number of Windstorms Causing Sufficient Damage for Holdings to Qualify for Insurance Benefits

	1962–63	1963–64	1964–65	1965–66	1966–67	1967–68
Dominica	20	15	23	23	16	20
St. Lucia	26	39	22	20	17	14
St. Vincent	16	25	16	13	14	19
Grenada	3	1	11	7	9	8

Sources: WINBAN 1964, 1968.

impressive, but thoroughly depressing, series of calamities does not give a full accounting of the environmental problems experienced. The listing of hurricanes and tropical storms in Table 2.1 is only partial. Also, less intense windstorms not on the list buffet the Windwards and often do considerable damage, especially given their greater frequency than hurricanes and tropical storms. Early WINBAN annual reports provide information on the frequency of all windstorms causing sufficient damage for affected holdings to qualify for insurance benefits (see Table 2.2).

Natural disasters are not the only environmental conditions affecting returns to banana farmers. The role of seasonal rainfall patterns also requires consideration. Fruit quality, a key determinant of prices, is usually highest during the driest part of the year, from January to May. The wetter, more humid conditions later in the year are more conducive to both leaf spot infestations and crown rot, a fungal disease that affects the crowns on the hands of the fruit and consequently leads to serious fruit quality problems.[2] Also, latex in the banana plant runs more freely at times of higher rainfall, contributing to latex staining of the peel of banana fingers, another quality defect. Farmers harvesting and packing bananas tend to work more hurriedly in the rain, leading to more bruising of fruit, a problem exacerbated by fruit becoming softer in the wet season. At this time, it is also difficult for trucks to travel over the slippery, unpaved feeder roads to reach many banana fields away from the main roads, thus necessitating the carrying of harvested fruit longer distances on people's heads, which further aggravates quality problems. And toppling of fruit-laden plants increases when the soil is saturated, as the shallow root systems fail to anchor the bananas adequately under such conditions, while the increase in the incidence of wet and dirty cardboard boxes, in which growers pack their fruit, leads to higher rejection rates.

Thus, environmental conditions cannot be ignored in the analysis of the Windwards industry and its degree of competitiveness. The regularity of the onslaught of environmental crises has left the industry, for much of its existence, in a perpetually weak financial condition. And the relationship of the Windwards to Geest only compounded these problems.

The World of Geest

In the early 1990s, Geest characterized itself as the largest importer and distributor of fresh produce in the United Kingdom (Roger Hilborne, pers. comm., 15 October 1990). The transnational corporation, the exclusive marketer of Windwards bananas from 1952 to 1995, had sales of £660 million in 1995, with interests not only in importing and distributing bananas and other produce but also in horticultural products and prepared foods. Until the sale of its banana business in late 1995, bananas were the centerpiece of the company and provided the basis for its growth and diversification into other areas (Trouillot 1988). It had been the largest importer of bananas into Britain from the 1970s until the first half of the 1990s.

Geest once owned banana-producing estates on both Dominica and St. Lucia, though following the trend of many other transnationals in the agricultural sector, it later terminated its involvement in direct production, thus removing itself from the riskiest and least profitable part of the banana commodity chain. It owned four refrigerated ships that carried most of the Windwards fruit on the nine-day voyage to the United Kingdom.[3] When reaching port in the United Kingdom, Geest sold some green fruit to the other two major importers, Fyffes and Jamaica Producers, and to other independent "green handlers." It transferred the majority of its imports—usually 55 to 65 percent in the latter half of the 1980s and early 1990s—to one of its ten ripening centers scattered throughout the United Kingdom. Here it ripened green bananas for a period usually from four to eight days at a carefully controlled temperatures of 61–63°F and at high levels of humidity (roughly 90 percent). After the bananas were ripened, its workers divided the fruit into two groups. The best grades were reserved for the large supermarkets or "multiples," such as Marks and Spencer and Sainsbury. Traditionally, its employees at the ripening centers cut the hands of such prime fruit into clusters, priced them for the multiples, and then sorted them according to color, as different customers preferred different shades of yellow in the bananas they received. The second group of fruit, the lower grades, received minimal processing, being shipped to banana wholesalers. As is true of the banana industry in general in the United Kingdom, the proportion of Geest fruit destined for the supermarkets has increased over time, a trend of particular significance for the Windwards. Supermarkets are much more discriminating concerning quality and, indeed, have frequently complained to Geest about the quality of Windwards bananas (WINBAN 1986a: 15)—adding pressure to improve fruit quality at the farm level.

The transnational's involvement in the Windwards industry has received a considerable amount of scrutiny in the literature, almost all of it from a negative perspective (for example, Marie 1979; Barry, Wood, and Preusch 1984; Thomson 1987; Thomas 1988; Trouillot 1988; Grossman 1994; Nurse and Sandiford 1995), with some exceptions (see Welch 1996). Much of the controversy involves the various contracts Geest has had with the Windwards, the imbalance in the distribution of wealth from the industry in favor of Geest, the disproportionate risks assumed by the Windwards, the issue of quality determination, and the intensifying pressures for technical change that emanated from the company—all classic areas of contention in contract farming.

Banana transnationals are certainly no strangers to controversy. The "big three" United States firms involved in the Latin American trade, as both producers and marketers, have developed unsavory reputations for their predatory commercial practices and involvement in Latin American politics (Burbach and Flynn 1980). Clearly, United Fruit's prior nickname as *El Pulpo*, or "the octopus" (George 1977: 144), was not meant as a term of endearment. As Thomson (1987: 14) points out, Geest is "hardly comparable to the infamous 'big three' companies who have traditionally dominated the industry in Latin America."

Similarly, Geest treated the Windwards better than Fyffes treated Jamaica. Geest never unilaterally terminated its contract with the Windwards as did Fyffes with Jamaica, and it tended to pay the Windwards slightly more for their bananas (Beckford 1967: 31). But as other observers (Marie 1979; Barry, Wood, and Preusch 1984; Thomas 1988; Thomson 1987; Trouillot 1988; Nurse and Sandiford 1995) have asserted, the company nonetheless reaped the majority of benefits from the banana trade.

Such an imbalance in the distribution of wealth in agricultural commodity systems is hardly novel, as anyone with even a slight familiarity with the development literature will realize. But what is unique is the way in which consciousness about the degree of exploitation had been diffused by the nature of the commodity chain and the mystification of relationships.

Geest did not have contracts with individual growers, following a tradition established by the Canadian Banana Company in the 1930s and later by Antilles Products Limited. Rather, Geest traditionally negotiated with WINBAN (and, more recently, WIBDECO), which functions on behalf of the four banana growers' associations, groups that were also signatories to the contracts. Positioning the associations between itself and the growers was a strategically astute decision. By not having to deal directly with growers, Geest spared itself the costs of providing extension services and credit to growers and the burden of managing and supervising the industry. More specifically, the relationship deflected the consciousness of most growers from the source of much of their problems. Farmers tended to blame the banana growers' associations, not Geest, for the prices received and for the burdens related to the constant round of technological innovations, many of which were the direct result of pressures from Geest.

Furthermore, the adversarial relationship between buyer and seller that characterizes many contract-farming systems was notably absent for much of the history of the Windward Islands' relationship with Geest (Thomson 1987: 35–36). One of the difficulties in bargaining with Geest was the dual perception of Geest as business partner and friend. For example, John Compton (1965: 20), former premier of St. Lucia, exclaimed "that we number among our blessings the fact that we have been fortunate to be dealing in this complicated banana business with a most vigorous, understanding and sagacious man in the person of Mr. John van Geest, head of our Marketing Company." At approximately the same time, the WINBAN (1966: 12) annual report for the year 1965–66 contained the following praise:

> It is, of course, well known to everyone by now that Mr. John van Geest, as far as the Company is concerned, is to the largest extent responsible for this satisfactory state of affairs through the personal kindness, interest in and disposition towards the Windwards banana industry. We would like, on behalf of the whole industry, to record our appreciation and thanks for all that he personally and the Company have meant to the Windwards.

Such sentiments can be found throughout official WINBAN and SVBGA documents.

Discussions about Geest and profits invariably turn to the issue of contracts with WINBAN and the Windwards associations. Since signing ten-year contracts with the SVBGA and the other associations in 1954, Geest had guaranteed its purchase of all bananas of exportable quality, a commitment that it kept through time. New contracts came into force in 1964, 1977, and 1995, with a multitude of additions and modifications to these agreements in the intervening years. Despite all the contractual modifications made up to 1995, when a substantial transformation in contractual relations finally occurred, a basic feature of the agreements remained to the benefit of Geest—the cost-plus nature of the contracts.[4] Geest had been allowed to deduct the costs its incurred in loading bananas in the Windwards, in shipping them to Britain, in unloading them in the United Kingdom, and eventually in its ripening and distributing activities. Allowance for a handling change guaranteed a profit. The cost-plus nature of the contracts insulated Geest from the risks, passing them on to the associations and, ultimately, the farmers.

Up until 1983, the basis for determining the price paid to the Windwards was the Green Market Price, the price at which Geest sold green Windwards bananas to "green handlers," other importers and ripeners of fruit. The formulation of the price was indeed somewhat complicated, as indicated by a WINBAN (1980b: 30) review of the pricing structure: "The Green Market Price is negotiated weekly between Geest and WINBAN. It is intended to reflect anticipated market conditions based on current market trends, as the price negotiated in any one week is in respect of fruit loaded for that week in the Windwards, but which will not reach the market until the following two weeks, taking into account voyage time and the ripening period." Once the Green Market Price was agreed upon, Geest then made deductions for its various costs.

The basic problem for the Windwards was determining what Geest's costs actually were. The ability of the Windwards to examine Geest's accounting procedures was limited compared with the sophistication of Geest's accountants. Complicating matters was the difficulty in ascertaining the accuracy of Geest's allocation of costs among its various enterprises. Thus, its shipping facilities in the Windwards served not only the banana industry but also its activities in importing and exporting other products from the islands. Even such seemingly minor issues, such as allocating the cost of lightbulbs, were potential sources of profit for Geest. WINBAN had also expressed concern over being charged for the full depreciation of the original costs of Geest's buildings in the West Indies without receiving any equity in the properties (WINBAN files). And charges for depreciation costs for Geest's ships proved equally questionable (Ecumenical Committee for Corporate Responsibility 1994). Potentially lucrative aspects of contracts for the Windwards failed to materialize. According to one of the contracts, the Windwards were supposed to share in the profits from Geest's ripening operations in the United Kingdom, but Geest claimed that the level of profitability in their ripening operations never achieved the level at which the Windwards were entitled to share in the returns (WINBAN 1980b: 25). In a 1980 review of the situation, WINBAN complained about the contractual relation: "The meeting recorded in various forms the views of persons present that WINBAN had insufficient knowledge of the methods used by Geest Industries in allocating costs which that Company exercised sole authority to budget and incur. Added to this, the financial data related to these were given to WINBAN at so short notice that it became difficult to attempt meaningful discussion on it" (1980b: 1). Documents recording WINBAN's concerns during contract negotiations in 1983 summarized the frustration (WINBAN files): "WINBAN claim that the Contract overprotects Geest which is able to recover all its costs irrespective of tonnage/price."

Increasingly concerned about the disappointing returns from the contracts, the Windwards finally achieved a significant change in 1983, the first time truly intense and contentious bargaining occurred between the parties. WINBAN demanded that the Windwards finally share in the returns from the sale of ripe fruit, not just green bananas. Consequently, a new basis for the price of Windwards bananas was established, the Green Wholesale Price, a weighted average based on both the traditional Green Market Price and the sale price of ripe fruit from Geest's operations. Moreover, a change, which was incorporated increasingly into subsequent contractual revisions, shifted a portion of the risks to Geest—some of Geest's allowable contractual charges were now based on a percentage of the Green Wholesale Price, not the actual costs claimed by Geest. These and subsequent modifications did improve the returns to growers somewhat.

But the fundamental imbalance in the distribution of rewards remained. Debates about appropriate and fair percentage charges now ensued, while certain fixed costs still remained in the contracts. Moreover, Geest could skim profits in other ways. For example, WINBAN grew increasingly concerned and suspicious about Geest's practice of deducting rebates allowed to its customers who supposedly were dissatisfied with and returned to Geest particular Windwards shipments of ripe fruit. The bizarre result was that in some cases the Green Market Price component (based on the sale of unripened fruit) of the Green Wholesale Price was actually higher, according to Geest, than the component based on the price (minus deductions and rebates) of ripe fruit!

Relations between the Windwards and Geest became increasingly strained in the early 1990s, in part because of controversies over contractual relations. But suspicions also arose in relation to Geest's maneuvers ahead of the implementation of the SEM. Without consulting the Windwards, it purchased 9,600 acres of land in Costa Rica in 1991 to grow its own bananas to market in Europe, essentially implementing a "multisourcing" strategy, which has become increasingly fashionable for transnational firms.

A change of historic proportions in contractual relations finally was achieved in the new agreement negotiated in 1994, which came into effect the following year. Ironically, the basis for the increased bargaining power of the Windwards was the same force of globalization now threatening to extinguish the industry—the Single European Market. Because of the regulations of the new banana regime, Geest now depended on its importation of Windwards bananas for access to lucrative dollar fruit licenses, and the Windwards thus demanded to share in those benefits. Consequently, the Windwards were able to alter the contract fundamentally. According to the new agreement, WIBDECO, the successor to WINBAN, was to assume responsibility for loading of fruit in the Caribbean and for unloading in the United Kingdom. Geest was contracted to provide shipping services for a period of five years, but WIBDECO became the official importer, a status that, according to EU regulations, gave it the right to obtain lucrative dollar licenses on its own in 1997, a financially significant benefit. Moreover, the Windwards guaranteed to sell only 78 percent of their bananas to Geest, with the remainder destined for Fyffes and Jamaica Producers. As a result of these changes, the Windwards were no longer affected by Geest's costs and deductions, as Geest was now merely the shipper, purchasing its fruit from the Windwards "free on truck" in the United Kingdom at prices determined by joint consultations among WIBDECO, Geest, Fyffes, and Jamaica Producers. In addition, both Geest and the Windwards were to benefit from the dollar licenses resulting from the importation of Windwards fruits.[5]

Thus, although Geest did not deal directly with growers, its various contracts were just as significant to them as agreements in other contract-farming schemes made directly between buyers/processors/marketers and farmers. Moreover, the subterfuge, manipulation, and subjectiveness involved in contract interpretations and implementations that plague many contract-farming schemes (Glover 1984) were also evident in the actions of Geest.

Clearly, for most of the history of the industry, the contracts ensured that the returns to the Windwards and their growers were marginal at best.[6] The 1974 SVBGA annual report summed up the issue concisely: "The banana industry, from its inception, has been beset by problems, the most serious of all being the poor return to the grower, who has never been able to eke out more than bare subsistence from the price that has been paid to him over the years" (SVBGA 1975: 3). The main exceptions to this observation occurred in the mid-1950s and in the second half of the 1980s. It is illustrative to examine the history of nominal or current prices received by Vincentian growers from 1964 to 1994; they appear to have stagnated up until 1973 but since then have grown considerably (Figure 2.1). But the impact of inflation also affects growers' real incomes, and if one adjusts the official (nominal) prices to account for inflation (real prices), the returns to growers really have not markedly changed during this period, except for a slight upturn in the second half of the 1980s.

The thorny issue of quality determination—perhaps the major complaint in contract-farming enterprises (Glover 1984)—also plagued relations with Geest. The extreme slide in quality that Geest claimed occurred from 1961 to 1968 appears suspicious, likely a reflection of Geest tightening specifications for the determination of the best grades. One of the basic problems is that historically the Windwards have had to rely on the honesty of Geest in quality assessments of ripe fruit, which were performed by the transnational in Britain.[7] Indeed, the Windwards have felt frustrated by the inability of all the technical innovations in production, harvesting, and packing to improve the level of quality to meet

FIGURE 2.1. Banana Prices, Current and Real, 1964–1994

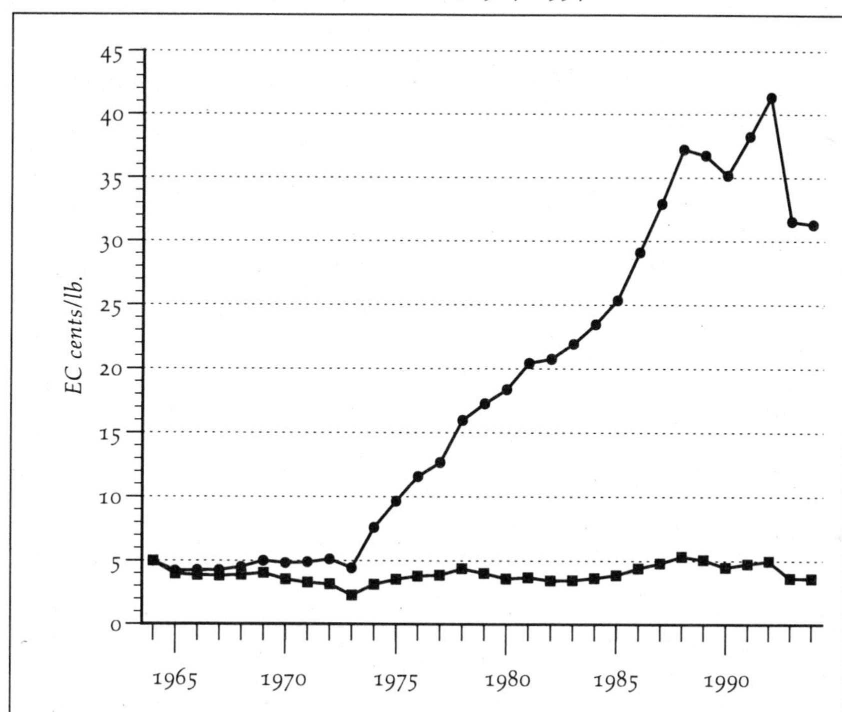

Sources: Felton 1981a; SVBGA annual reports; SVG Statistical Unit digests of statistics and files.

the ever-increasing demands of Geest. For example, the Windwards complained that, after all the costs associated with the transition to boxing plants, quality did not improve in the 1970s as expected. The SVBGA (1972: 2) lamented that "what baffles us is that in spite of all the care taken and effort made by our farmers and boxing plant personnel, we continue to receive very unsatisfactory and unfavourable out-turn reports." Similarly, WINBAN officials (WINBAN 1973: 8) declared: "The assessment results which were being sent to us from Spalding [where Geest did assessments] were still very mystifying. Fruit which was confidently expected to produce a much higher out-turn grade far too frequently obtained low marks. This was most discouraging."

A new form of packing bananas called "field packing" (see Chapter 5) introduced in the 1980s initially brought significant improvements in fruit quality (as measured by the percentage of fruit in the highest two grades)—and hence returns to growers (see Figure 2.2). But then, according to Geest, quality started to fade as the incidence of crown rot supposedly increased; fungicide-treated

FIGURE 2.2. Banana Quality, St. Vincent, 1983–1996

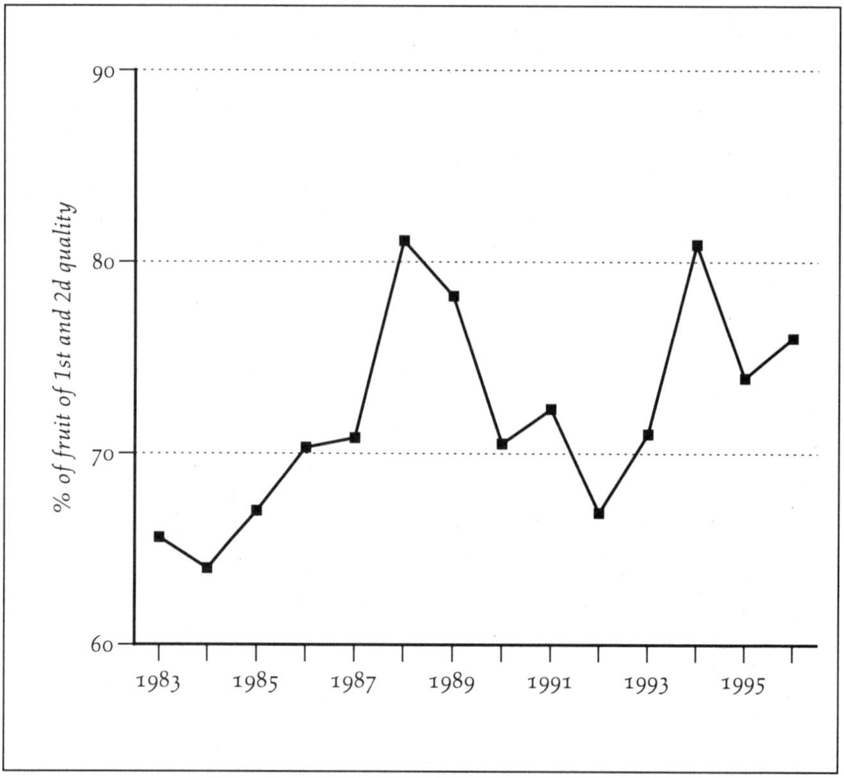

% of fruit of 1st and 2d quality

90
80
70
60

1983 1985 1987 1989 1991 1993 1995

Source: SVBGA files and annual reports.

crown pads that were used in field packing and were initially effective in controlling the problem mysteriously lost their effectiveness a few years later. Interestingly, increasing complaints by Geest about problems concerning crown pads and crown rot came at a time when it was trying to convince the Windwards to adopt a new method of packing bananas that it believed would enhance the marketability of fruit while also reducing its own costs—one that, incidentally, also required the adoption of alternative means of controlling crown rot.

Certainly, the Windwards have had quality problems with their fruit and Geest has not been the only force pressing for improvements in quality; the British government's policy of setting quality targets for the Windwards to achieve in the 1980s—with the threat that the quota on Latin American fruit would be increased if the targets were not met—was another significant influence. But the assumption made by Geest was that quality problems were always the fault of the Windwards and that once the company took control of the bananas, the utmost care was exercised to handle and process the fruit properly.

Whatever problems happened in relation to refrigeration, handling, and ripening under the control of Geest were never known to the Windwards—but such problems inevitably negatively affected returns because of the nature of the contracts.[8]

A key role played by capital in contract-farming systems is the influencing of production patterns, and Geest was instrumental in this process. Although WINBAN had devised some innovations on its own and others resulted from WINBAN-Geest collaborative efforts, the transnational had clearly been the driving force behind most technical innovations designed to improve marketability and quality, often exerting pressure to induce change in Windwards practices. While both Geest and the Windwards benefited from better-quality fruit, it was the growers who had to bear the burden of implementing the changes. A report by WINBAN (1971: 3) is illustrative: "The demands requested of us by our Marketing Agent, Messrs. Geest Industries, over the past two years through our local Association, for the better handling of our fruit with the objective of IMPROVING THE QUALITY OF OUR BANANAS, has cost the farmer a considerable sum for the purchase of foam sheets and diothene tubing for the protection of the fruit." Much the same can be said of many other technological innovations. The recent change to yet another form of harvesting and packing, called the "cluster pack" method, was clearly a response to Geest pressure.

The Windwards did sometimes resist such demands. For example, WINBAN refused Geest's urgings in the early 1990s to force all Windwards growers to place brand identification labels on each hand of fruit while in the field because WINBAN felt that it would be too time consuming, a practice, incidentally, that would have saved Geest labor costs in its United Kingdom operations. But in most cases, Geest was able to achieve its objectives, either by subtle persuasion and cooperation or by considerable pressure.

But Geest will be making no more demands on the growers. It decided to sell its banana business to a joint WIBDECO-Fyffes venture in late 1995.[9] Geest lost substantial amounts of money in its own Costa Rican operations, affected in part by an outbreak of black Sigatoka, a fungal disease. It found that it was much easier to cajole the Windwards into producing better fruit quality and increasing productivity than to perform the feat itself. Furthermore, a bitter strike broke out at its plantation in 1994, in which eighteen workers were shot and three seriously wounded. Coupled with receiving lower production than expected from the Windwards and increasing competition and weak banana prices in the SEM, events convinced Geest to terminate its involvement in the trade. But perhaps another unstated reason was equally significant. The basis of Geest's profitability in the banana industry had always been the inequitable contracts that it had with the Windwards. With a more equitable contract finally instituted in 1995, a change facilitated by the process of globalization, as represented by the SEM, Geest perhaps realized that the key source of its profitability in the trade had finally disappeared.

British Involvement in the Vincentian Banana Industry: The Case of Aid

This review of the negative impacts of the environment and capital that impinge on the Windwards banana industry is not solely an attempt to document the harsh reality of pressures that contract farmers face. Rather, it is also intended to raise the crucial issue of how the Windward Islands banana industry managed to survive in the face of such severe and constant adversity. My focus on explaining the survival of the banana industry provides a central role for the state. In this case, British foreign aid was the crucial pillar of support that ensured survival of the industry through repeated crises (Grossman 1994). Certainly, the preferential system with its quota on Latin American fruit was also essential—a fact well recognized in the literature (see Marie 1979; Rojas 1984; Thomson 1987; Trouillot 1988). But the role of British foreign aid—which was substantial in amount and provided on a fairly regular basis—has been ignored in the literature on the industry. Without continual infusions of aid, the perpetually financially weak banana industry would most certainly have collapsed, irrespective of the protection provided by the British preferential system of tariffs and quotas (Grossman 1994). Moreover, the nature of aid has affected the labor process and technological change in contract farming.

The British rationale for the numerous programs of aid can be examined on two levels. The first involves trying to solve the immediate crises that the Windwards industry faced—and there were certainly many economic and environmental crises. Aid programs also focused on encouraging farmers to use more agrochemicals and adopt new production and harvesting techniques, with the expectation that farm productivity and fruit quality would improve.

The second level involves inquiring why the British wanted to continue supporting the industry. Although concern for the welfare of the Windwards was significant, the primary reason rests with the interests of the United Kingdom itself, which have changed over time. Initially, aid was allocated to help establish an industry that was needed to provide a sterling source of bananas for the undersupplied U.K. market in the 1950s in the hopes of alleviating Britain's balance of payments problem. Continuing concerns about political stability in the region were also important (Thomson 1987). Indeed, John Compton highlighted the positive influence of the industry on stability, contrasting the situation there with the turmoil elsewhere in the Caribbean. Proceedings from negotiations concerning the future of British aid to the banana industry captured his sentiments (Windward Islands Banana Conference 1965: 4–5):

> In recent times they had seen chaos and upheaval in the West Indies. He [Compton] had only to mention Cuba, Haiti, the Dominican Republic and British Guyana to remind them of this. These territories were in close proximity to the Windwards. However, against this background of fragmentation and social disorder in the West Indies, the Windward Islands had been moving closer together in harmony. This was essentially due to the advent of the banana industry and to the establishment of the Windward Islands Banana Growers' Association, both of which had acted as a unifying force.

The motivation for aid to the banana industry soon shifted to making the Windwards more competitive with Latin American producers, so that eventually the British preferential market could be dismantled and the Windwards could compete in an open, unregulated market (ibid.: 2)—a goal that has never been achieved. By the mid-1970s, a somewhat different perspective was emerging. Aid was now predicated on the belief that there was no alternative besides the banana industry for the Windwards. A British Development Division (BDD) report asserted, "The industry is their only agricultural activity which can be sustained on any scale as the opportunities for other investment are limited by natural conditions or limited markets" (1975: 4). At the same time, the British realized that the failure of the banana industry would not only have serious implications for the Windwards but also for the United Kingdom itself:

> If the banana industry failed, the [Windwards] Governments (already finding it difficult to finance essential services) would certainly turn to HMG for significantly increased budgetary aid, which we should find it difficult, if not impossible, to refuse—and it might, be of indefinite duration. Against this background it is in our own interests, as well as those of the islands (and particularly of the smaller and poorer farmers) to explore urgently all possible ways of helping to make the industry more efficient. [5]

Certainly, other considerations entered into state policy. Geest depended on the Windwards banana industry for its source of wealth, and Geest was, after all, a British corporation. Similarly, funding for the Windwards was usually "tied" aid, requiring expenditure of British funds on British products, especially agrochemicals. But these considerations were certainly secondary in nature, with concerns for the British state's own initiatives primary.

We can examine aid to the SVBGA as an example of aid to the Windwards industry generally.[10] In some cases, aid to St. Vincent was part of broader

TABLE 2.3. British Aid to the St. Vincent Banana Growers' Association

Date	Program	Amount (EC$)
1954–56	Grants/loans to establish SVBGA	212,000
1955–65	Price Adjustment Scheme	177,000[a]
1963	Hurricane and leaf spot relief	229,000
1968–71	Replanting Incentive Scheme	—
1973	"Shot-in-the-Arm" agrochemical subsidies	528,000
1974–76	Marriaqua Replanting and Rehabilitation Scheme	456,000
1974–76	Fertilizer Subsidy Scheme	1,172,000
1975–76	Borer Control Program	286,000
1976–77	Interim Banana Development Program	381,000
1977–82	Banana Development Program	3,135,000
1979	Volcanic eruption rehabilitation	2,034,000
1980	Hurricane Allen rehabilitation	2,455,000
1982	Emergency fertilizer	330,000
1983–86	Banana Industry Support Scheme	1,480,000

Source: Grossman 1994.
[a]Estimate

schemes encompassing all the Windwards. In others, British aid was intended specifically to ameliorate the devastating impacts of natural disasters affecting St. Vincent, though the other Windwards received similar forms of assistance when they, in turn, were plagued by such disasters (Table 2.3).

St. Vincent and the other Windwards benefited from the Price Adjustment Scheme, which lasted from 1955 to 1965, in which the British government supported the price of fruit when it fell below a certain target level to bring stability to the market and provide encouragement to growers to expand output. St. Vincent also received significant amounts of British aid after environmental disasters and pest infestations (1963, 1975–76, 1979, and 1980) and after increased fertilizer costs or declining banana revenues limited the Association's ability to afford the importation of adequate agrochemical supplies (1973, 1974–76, 1982). In such cases, aid was largely in the form of agrochemicals, primarily fertilizers but pesticides as well. St. Vincent was also involved in numerous British-funded replanting and rehabilitation schemes, which were generally Windwards-wide: the Replanting Incentive Scheme (1968–71), the Marriaqua Replanting and Rehabilitation Scheme (1974–76), the Interim Banana Development Program (1976–77), the Banana Development Program (1977–82), and the Banana Industry Support Scheme (1983–86). The general purpose of these programs was to enhance extension services; to encourage farmers to improve current practices and adopt new production, harvesting, and packing techniques and to use agrochemicals more intensively; to raise the level of fruit quality; and to expand exports. Provision of free and subsidized agrochemical inputs was also a feature of most programs.

These replanting and rehabilitation schemes were all inaugurated with the most optimistic of outlooks. The Marriaqua program was supposed to "lay the foundations for a viable and competitive banana industry through achievement of more regular supplies, higher productivity, and better quality" (SVBGA files), and the Banana Development Program, the most ambitious of them all, costing the British government EC$16,719,000 for the entire Windwards-wide, five-year program, was intended to make the industry "self-supportive and competitive on the world market by 1982" (BDD files). Yet the aid received never enabled the Windwards to achieve these goals.

The massive amounts of aid poured into the industry have certainly achieved some of the desired results. Assistance during economic and environmental crises has been essential in keeping the industry afloat. Aid programs strengthened the extension services. The Banana Development Program led to improvements in cultural practices and fruit quality, the latter resulting from the introduction of field packing. Aid programs have also resulted in farmers using more agrochemicals, which has boosted productivity; yields on St. Vincent, for example, have risen from three to four tons per acre in the 1960s to six to eight tons in the late 1980s. More generally, aid has had a positive impact on the economy and welfare in the islands by ensuring the survival of the industry; the demise of the industry, which would have been inevitable without financial aid, would have been particularly devastating for the thousands of farmers and their workers who depend on banana exports for their livelihoods.

And dependence on aid continues. As a result of the increased competitive pressures in the SEM and the declining value of the British pound (see Figure 1.3), the industry once again found itself in a severe financial crisis. In the current context, aid now comes primarily from the European Union in the form of Stabex funds; such assistance has been part of the Lomé Conventions and is intended to help ACP countries that depend on a small number of exports to cope with sudden drops in export earnings caused by fluctuations in world prices or sharp variations in output resulting from environmental crises. In 1994, St. Vincent applied for EC$15,487,892 to fund its Banana Industry Development Programme, with the goal "to revitalise and sustain banana production and marketing in a competitive environment" (SVBGA 1995: 6). This replanting and rehabilitation program—with its familiar but always elusive goal—thus joins a long series of previous projects designed to rescue the industry.

This analysis of aid has focused on the peasant-state relationship, but the significance of aid has been much broader. In particular, we must also examine the benefits to capital. Specifically, Geest also profited from British aid, which saved the transnational money by enabling it to pay the Windwards less for bananas than was necessary to ensure the survival of the industry (Grossman 1994). British aid was always available to supplement the meager earnings of the Windwards associations and keep them functioning during repeated crises. Without continual infusions of British aid, Geest would have had to pay the Windwards a larger percentage of its revenues from the sale of bananas, or the industry would have failed. For those familiar with the development literature, such a subsidy to Geest seems and is in fact strikingly similar in structure to the widely recognized subsidy that subsistence agriculture provides to capital; researchers note that because peasants engage in subsistence production and thus partly supply their own household needs, capital can pay them less for their crops and labor because the peasants are not fully dependent on their cash earnings for survival. Indeed, Trouillot (1988) appropriately makes this precise point concerning subsistence production and Geest's relationship to banana farmers on Dominica. In essence, in the case of the banana industry, Geest benefited from a double subsidy—from subsistence agriculture and from British aid—which helped to make its relationship with the Windwards a very profitable one and the basis for its expansion into other endeavors.

The St. Vincent Banana Growers' Association and the Vincentian Banana Industry

It is a warm Saturday evening in St. Vincent. The time, 7:10 P.M. Listeners to the national Vincentian radio station are treated to the song "Banana Man," the introduction to the weekly, twenty-minute program "Banana Corner" produced by the SVBGA. Into this short time slot the show's announcer packs a plethora of information about export statistics to date, the grade of bananas to be cut for the upcoming weekly harvest, forthcoming local meetings, and recent fruit quality data. The hurried pace of the show is then punctuated by another popular song, "We Love Bananas." Finally comes the heart of the show, a segment discussing appropriate production and harvesting practices—topics such as how to reduce leaf spot problems, the proper control of nematodes, the advantages and disadvantages of sleeving banana bunches, and how to improve field sanitation. Occasionally, a recipe incorporating bananas is added near the end. Then the announcer reminds his audience, "Quality is crucial. Leave bad bananas is your field."

This program is only one of the many ways in which the SVBGA impinges on the lives of Vincentian banana growers.[11] This organization performs the classic functions of the central buyer/coordinator in contract-farming enterprises. Formed initially in late 1953 as a public company with shares and transformed into a statutory corporation by the Banana Growers Association Ordinance, 1954 (SVG 1955), it provides banana growers with extension services, inputs on credit, and pesticide control services, as well as a guaranteed market for their crop.

Although it has no formal, written contracts with individual farmers, in contrast to the situation in most contract-farming systems, the de facto situation is the same because, according to its charter, it must purchase all bananas of exportable quality. Moreover, regulations incorporated into various government statutes authorize the SVBGA to supervise and control the production process. For example, the Banana (Protection and Quality Control) Act, 1984 (SVG 1984) specifies that "[a]ll exportable bananas shall be grown, harvested, processed, packed, handled and otherwise dealt with and transported in the prescribed manner" and that "[i]n order to ensure that proper disease and pest control are maintained, field sanitation shall be of a high standard, in that proper weed control, proper drainage and recommended plant density shall be maintained and, dead and drying leaves shall be cut-off." Other regulations in the act pertain to additional aspects of cultivation and proper procedures for harvesting and packing bananas, as well as to criteria for rejection of fruit. The statute also provides for fines of up to five hundred dollars for violations of the regulations. The intention of the act is to establish WINBAN's recommendations as the "prescribed manner." Although these recommendations are not highly detailed in the act and few, if any, growers ever received fines under the law in St. Vincent, the regulations do indicate the legal basis for intervention by the state in the production process.

Although the SVBGA always provided a guaranteed market for the crop, had its own extension service, and provided inputs on credit, whether it instituted a pure form of contract farming from the beginning in the 1950s is open to debate; adoption of specific cultivation and harvesting procedures was really encouraged through the payment of bonuses rather than by direct regulation, giving farmers flexibility in their production patterns. But by the 1980s, when field packing was introduced, farmers had to harvest and pack their bananas in a very specific manner or their fruit would be rejected. Thus, by this time, control over at least a part of the production process was clearly evident, and given the guaranteed market that had long been established, the system certainly satisfied the two basic criteria of contract farming.

Membership in the SVBGA is relatively easy to obtain because, unlike in some contract-farming schemes, the number of growers is not limited. To become a registered member of the SVBGA, a person need only complete a simple, one-

TABLE 2.4. Number of Growers, by Amounts Sold, 1992

Amount Sold (tons)	Number of Growers
<1	1,201
1–2	787
2–3	682
3–5	1,077
5–7	781
7–10	937
10–15	941
15–20	524
20–25	289
25–30	172
30–40	227
40–50	107
50–80	89
80–100	15
>100	26
Total	7,855

Source: SVBGA files.

page application form, be at least sixteen years old, and, according to statute (Banana Industry Act, 1978 [SVG 1978a]), have planted at least thirty bananas, a minimal requirement, as that amount would necessitate a holding of less than one-twentieth of an acre. More than one person from the same household can register, a recognition of the pattern of land tenure in which rights vest in individuals, not households.

Even though each grower must register with the SVBGA, the exact number of active growers is difficult to determine. The Association's statistics (SVBGA files) indicate that there were 12,695 registered growers in 1992, but only 7,855 actually sold bananas that year (Table 2.4). But the actual number of "real" growers is difficult to determine because of the problem of "multiple registration," a practice that the SVBGA views as insidious but whose frequency is unknown. Multiple registration occurs when an individual grower obtains one membership in his or her own name and a second one in the name of a different household member, relative, or even fictitious individual, with such farmers selling their own fruit under both names. Some engage in the practice to avoid repaying credit from the SVBGA or loans from banks. But others have a non-malevolent intent. Peasants prefer to keep records of their banana sales from land that they sharecrop separate from those dealing with their plots held under other forms of tenure to avoid possible conflicts over amounts owed to owners of the sharecropped parcels.

TABLE 2.5. Ratio of Extension Agents to Active Banana Growers, St. Vincent, 1987–1994

Year	Ratio
1987	1:461
1988	1:653
1989	1:564
1990	1:717
1991	1:557
1992	1:561
1993	1:628
1994	1:558

Source: SVBGA 1996a: 3.

Despite problems involved in determining the precise number of growers, peasant cultivators with banana holdings of five acres or less predominate in the industry. Employing the approximate yield of six tons per acre relevant to 1992, the data in Table 2.4 suggest that approximately 94 percent of the growers cultivated five acres of bananas or less.[12] Furthermore, the "large-scale" banana plantations found in Latin American, where properties of several thousand acres are not uncommon, are clearly not evident in St. Vincent and the rest of the Windwards, owing to the decline of the estate sector in the twentieth century (see Chapter 3). In the late 1980s, the largest banana holding on St. Vincent, owned by the Vincentian government, was only 112 acres, with the largest private banana property being only 93 acres. Since then, both have been subdivided, and by 1995, the banana holding of greatest size on the island was approximately only 60 to 70 acres.

The large number of small-scale growers has implications for the functioning of contract-farming enterprises, especially in relation to extension services. One of the keys to successful contract farming is a high ratio of extension agents to farmers. Given that banana production is much more technologically complex than any other system of agriculture ever practiced on St. Vincent and that the pace of technological change in the industry has accelerated over time, a high ratio of extension agents to farmers would be expected. That has not been the case, however, as data from 1987 to 1994 illustrate (see Table 2.5). In contrast, many contract-farming schemes involving labor-intensive crops in sub-Saharan Africa have much higher ratios of 1:50 to 1:200 (see Watts et al. 1988: 178). For example, in the successful tobacco contract-farming scheme operated by British American Tobacco in western Kenya, the high ratio of 1:50 enabled extension officers to visit each grower once every two weeks to supervise their activities; in comparison, many Vincentian banana growers do not see their extension agents more than once every one or two months or even longer. The inability of the SVBGA to provide adequate extension services stems directly from its usually weak financial condition, which reflects the combination of environmental, political-economic, and market forces that have constantly assaulted the industry.

The implications of such insufficient extension assistance in the context of complex and rapid technological change are profound. First, the ability to monitor and regulate the behavior of farmers and to induce uniformity in production practices is limited—a condition that is highly relevant to the issue of the deskilling of labor. Second, the ability to educate farmers adequately and foster improvements in fruit quality, one of the keys to competitive success, has also been constrained.

The SVBGA provides other services as well, many of which are characteristic of buyers/central coordinators in contract-farming schemes.[13] It is the island's largest importer of agrochemicals—bringing in over ten thousand tons of fertilizers annually in the late 1980s. And, benefiting from a duty-free concession on such imports, it sells agrochemicals to farmers on a nonprofit basis, below market rates. To encourage farmers to use fertilizers and pesticides, the SVBGA has, since 1974, withheld a "cess," or deduction, from each payment made to growers; farmers are supposed to use the accumulating cess account funds to purchase agrochemicals every two months.[14] The Association also provides credit for obtaining additional inputs, charging no interest and requiring a 50 percent deposit on fertilizers and a 25 percent deposit on other inputs. It is responsible for leaf spot control, utilizing ground crews and aerial spraying to combat infestations. In addition, the SVBGA organizes and provides transportation for shipping of harvested, boxed bananas from countryside to inspection stations and town, having almost nine hundred truckers under contract to provide the service in the late 1980s and early 1990s.

As is characteristic of buyers/coordinators, the SVBGA has responsibility for the determination of fruit quality. It inspects most bananas today at the wharf in Kingstown, where on Mondays and Tuesdays long lines of trucks of various sizes clog the streets waiting for their cargoes to be examined. Previously, the Association paid truckers to bring fruit for inspection to the numerous Internal Buying Depots (IBDs), large warehouselike, galvanized-roofed structures scattered throughout the island.[15]

Whereas some buyers/central coordinators dishonestly change grading standards according to the availability of supplies or market prices to maximize profits (Glover and Kusterer 1990), the Association, which is a nonprofit organization, does not. Indeed, rejection rates in the late 1980s and early 1990s,

TABLE 2.6. Fruit Quality Defects Specified, 1993

1. Superficial bruises	14. Pest damage
2. Crown trimming	15. Residue
3. Damaged pedicels	16. Red rust damage
4. Dirty	17. Ripe and turning
5. Finger end rot	18. Scars
6. Flower thrips	19. Scruffy
7. Fused fingers	20. Short finger
8. Latex staining	21. Sooty mold
9. Maturity stain	22. Speckling/pin spotting
10. Misshapen/deformed fingers/units	23. Stale fruit
11. Mutilated fingers	24. Undeflowered
12. Overgrade	25. Undergrade
13. Peel burn	

Source: SVBGA 1993b.

which rarely exceeded 4 or 5 percent of fruit inspected, were low compared with those of many other schemes. Nonetheless, the criteria for grading and the number of defects that can lead to downgrading or rejection of fruits can be daunting. Certainly, consumers in the United Kingdom have no conception of the rigorous and complex criteria used to determine the acceptability of fruit that ultimately reaches their households. In the mid-1990s, a total of twenty-five different defects related to the characteristics of fruit could lead to downgrading or rejection (Table 2.6). In addition to those items on the list, other defects related to the boxing of fruit are grounds for rejection. Thus, even though bananas packed in a cardboard box are of the highest quality, if the container is dirty or wet, it will not be suitable for export. Similarly, a box that is poorly packed or underweight will not be accepted for purchase.

To encourage the delivery of high-quality fruit, the SVBGA pays farmers based on a scale according to the grade of fruit delivered, a pattern characteristic in contract-farming schemes. Actual payment schedules have varied over time. In the late 1980s and early 1990s, the SVBGA specified only two exportable grades, with the price difference between them of just EC$0.03 per pound. In 1994, the Association changed the system to include three grades and by 1996 had introduced five different classes with even more extreme price gradients, reflecting the SVBGA's desperate need to coax better quality from the growers, given the financial difficulties of the industry and increasing pressures from the EU (Table 2.7). Such detailed pricing scales are not uncommon in contract farming; an extreme example is that employed by British American Tobacco in Kenya, which lists fifteen grades, with the top and bottom prices differing by a factor of approximately six (Shipton 1985: 299).

TABLE 2.7. SVBGA Pricing Structures, 1994 and 1996

1994	1996
Basic price = EC$0.153/lb.	Basic price = EC$0.095/lb.
75–84% UWS[a] = EC$0.223/lb.	75–79% UWS = EC$0.2175/lb.
85–100% UWS = EC$0.273/lb.	80–84% UWS = EC$0.235/lb.
	85–90% UWS = EC$0.27/lb.
	91–100% UWS = EC$0.2875/lb.

Source: SVBGA files.
[a]UWS = unit within specification, which means either no noticeable defects or relatively minor flaws in relation to certain criteria (see Table 2.6). Scores are based on the percentage of all clusters within a small number of boxes sampled by the SVBGA.

The SVBGA is not independent in setting banana prices, being subject to Vincentian government influence in a variety of areas. The government must approve the SVBGA budget, price changes, and appointments of senior personnel. It also ensures that the activities of the thirteen members of the board of directors of the SVBGA, who establish policies for SVBGA management to follow, are consistent with government policy. Although the government, according to law, appoints only six members of the board of directors, while growers elect the other seven, the government, not the growers, controls Association policy; it also appoints the chair, who is one of the seven elected directors, selecting someone whose views are consistent with national policy.[16] The Vincentian government has the power to dismiss the board of directors as well, as it did in 1958 as a result of financial mismanagement and political activity (see Commission of Enquiry 1959). But in the management of day-to-day affairs and issues related to production goals and technological innovations, the SVBGA has considerable independence.

Government influence over and interest in the SVBGA reflects the overwhelming importance of the banana industry to the Vincentian economy and to the state budget. The government derives income from bananas both directly and indirectly. It receives income directly through the levy of an export tax on bananas, traditionally set at 3 percent but reduced to 2 percent in 1994. Indirectly, as the health of the banana industry greatly impacts the overall level of economic activity on the island, it also affects the general level of private spending and consequently influences the collection of import tax and consumption tax revenues crucial to the government budget.

But unfortunately for government revenues and the Vincentian population generally, conditions in the banana industry on St. Vincent, as elsewhere in the Windwards, have deteriorated significantly, starting in late 1992. Exports declined by 23 percent from 1992 to 1993, in spite of good weather. The SVBGA (1994: 8) lamented, "During the year, many fields were abandoned, while others received just the bare minimum cultural practices." A further decline of 47 percent in 1994 left export levels at their lowest since 1983 (SVBGA 1995: 1). The following year they rebounded 61 percent, but output was still far below that during the period before formation of the SEM; and recovery stalled again in 1996, when exports fell by 12 percent (WIBDECO 1997: 1). The crisis was evident in the SVBGA budget. During the latter half of the 1980s, the Association built up considerable financial reserves, but from a surplus of EC$10.73 million in 1991, the SVBGA developed a deficit of EC$8.27 million in 1992, further ballooning to slightly over EC$20 million each year from 1993 to 1995 (SVBGA 1993a, 1996a).[17]

The recent decline in the financial condition of the SVBGA and the fall in prices have simply reinforced the long-held disdain many villagers have for the Association. While the Association believes it is oriented toward helping small farmers, peasants' perceptions of the statutory corporation are quite different, as numerous villagers' comments indicate:

> It is ridiculous. Association never loses. Growers always working and always losing.
>
> They [SVBGA] have so much kind of scheming in these kinds of places.
>
> We only work to support people at the Association.
>
> The Association does tief [steal].
>
> If a box of bananas weighs 30½ pounds, we get credit for only 30, and they keep the rest.

In reality, Vincentian farmers tend to be suspicious about any large organization, and their views about the SVBGA are no exception. Many peasants complain that the Association listens to and helps only large-scale farmers, while ignoring their needs.[18] Rumors at the village level swirl about the inequities in the institution, including the belief that large-scale growers receive more lenient treatment in relation to credit and repayments. Indeed, some farmers have become so skeptical of the intentions of the SVBGA that they mimic the planting patterns of members of the board of directors in the belief that the board will raise prices only when their own fields are ready to be harvested.

Placing the SVBGA in perspective, however, requires considering the institution in light of other statutory corporations as well as transnationals involved in contract farming (see Glover 1984; Glover and Kusterer 1990; Little and Watts

1994). From this broader perspective, the SVBGA can be viewed more favorably. The Association's policies, which draw so much hostility from growers, in reality reflect the prices it receives and the pressures from Geest, the British government, British supermarkets, and the EU for better-quality fruit; although transnationals are usually the recipients of such ire, Geest had escaped such venom by dealing only with the associations and not directly with the growers. Also, the rampant corruption that has been a problem for statutory corporations in many developing countries has not, except for a few minor instances, been a significant or enduring feature of the organization. The Association has shown concern for the welfare of all growers after environmental disasters; to help farmers recover financially from losses attributable to hurricanes and tropical storms and the volcanic eruption in 1979, the Association has paid to affected growers short-term living allowances and compensation for damage to their banana plants, a practice that is certainly rare in contract-farming schemes. Furthermore, the SVBGA regularly pays growers on time, one week after delivering their bananas, a feature villagers take for granted. Yet, long-delayed payments to growers have plagued other statutory corporations, especially in sub-Saharan Africa. Also, the SVBGA faithfully fulfills its obligation of providing a guaranteed market; even when it decides not to export all bananas harvested to prevent oversupplying the British market, it will still purchase them from the growers, selling those not shipped to traffickers (Vincentians who export commodities to nearby islands) or giving them to government institutions, such as the hospital. Breaches of the commitment to provide a guaranteed market have certainly beset other contract-farming enterprises (Glover and Kusterer 1990).

In spite of the results of this more positive assessment, the SVBGA and other banana growers' associations are undergoing significant restructuring. A recent study of the industry by Cargill Technical Services of Britain (1995), endorsed by the EU and the Windward Islands governments, calls for replacing the boards of directors and senior management of all the associations because of financial and administrative mismanagement—a move most peasants would likely endorse.

The associations have been forced by the EU to accept these recommendations as a condition for future aid, which reflects the growing importance of the forces of globalization in the industry and in the lives of growers. But while the nature of the forces of globalization are new, external pressures and events have always shaped the lives of Vincentians and other peoples of the Caribbean. Such influences are evident in an examination of the historical development of the Vincentian economy, the analysis of which is crucial for understanding contract farming in the contemporary era.

Three

St. Vincent: Contemporary and Historical Perspectives

Visiting British official E. F. L. Wood (1922: 38) observed in the 1920s: "The British West Indies naturally rely for their economic prosperity on the export of tropical agricultural products. Their material well-being depends almost entirely upon obtaining overseas markets for their principal export crops, namely, sugar, coconuts, cocoa, bananas, Sea Island cotton, limes, nutmegs, and arrowroot." Such a pattern was, in fact, a continuation of a long-standing trend. Since its incorporation into the British colonial orbit in the eighteenth century, the Vincentian economy has been dominated by export agriculture. Consequently, the prominence of banana export production today follows the trajectory of Vincentian history. But the nature of contract farming in the banana industry is qualitatively different from other commodity systems utilized on St. Vincent previously. Thus, the role of capital and the state in the production process, agriculture's relation to the environment, the nature of markets, and the character of global forces impinging on the Vincentian economy have all changed significantly. At the same time, key dimensions of the Vincentian production process today, especially the importance of local food production and marketing and the unequal distribution of wealth, represent fundamental continuities with the past that are crucial for understanding the political ecology of banana production. Specifically, the nature of peasant agriculture has roots extending far back in time, having evolved from traditions established during the era of slavery and conditioned subsequently by interactions with the environment, the Vincentian state, estate agriculture, and foreign markets.[1] Furthermore, the long-entrenched tradition of struggle and resistance against estate domination during slavery and the postemancipation period has significant implications for the ability of capital and the state to control the peasant labor process in contract farming today.

Contemporary St. Vincent

Located in the southern part of the Lesser Antilles chain, the rugged, volcanic island of St. Vincent is only eleven miles wide and eighteen miles long (Map 2). Encompassing 134 square miles of land, it is the largest island in the Eastern Caribbean nation of St. Vincent and the Grenadines, the latter containing thirty-two small islands and cays scattered to the south, stretching toward Grenada. The economic hub of St. Vincent is Kingstown, the capital, which is located on the southern end of the elliptically shaped island, with a sheltered bay on one side and surrounded by verdant, rugged hills on the other. The city, the only true urban center on the island, and its suburbs contained slightly over 26,200 of the nation's 106,499 inhabitants in 1991 (SVG Statistical Office 1993), though several small, semiurban centers dot the windward (Georgetown, Calliaqua, and Mesopotamia) and leeward (Layou, Barrouallie, and Chateaubelair) sides. The capital—simply known locally as "town"—is laid out primarily on three streets parallel to the bay and is the center of government, finance, and commerce. It contains a mix of stone-faced, colonial-period buildings, vernacular architecture, and contemporary structures. Each day a steady stream of cars, colorfully painted passenger vans, and pickup trucks from throughout the island descend on the capital, linking rural and urban areas.

Evidence of the open nature of the economy, a characteristic that St. Vincent shares with its island neighbors as well as small countries in general (Streeton 1993), abounds in Kingstown. The port is regularly unloading cargoes ranging from construction materials to vehicles to household goods, with the majority of imports coming from the United States, though Trinidad and Tobago and Britain are also important suppliers. The numerous supermarkets contain a wide variety of imported foods that are popular fare and are the mainstay of Vincentian diets. The large warehouse of the SVBGA, filled with the acrid smell of packaged agrochemicals, continually dispenses loads of imported fertilizers and pesticides destined for farmers' fields. And most of the banks are branches of foreign corporations.

Indeed, the heavy import dependence of St. Vincent has led to growing visible trade deficits—another characteristic shared with other Caribbean islands. Thus, the trade deficit has ballooned from EC$23,443,000 in 1970 to a record EC$206,694,000 in 1993 (Figure 3.1) (SVG Statistical Unit 1995: 54). The actual impact of the trade deficits has been muted by remittances from overseas workers, a growing tourist sector, and large inflows of foreign aid (SVG Central Planning Division 1992: 1). Foreign aid has been particularly important not only for the banana industry but also for the functioning of the entire economy,

FIGURE 3.1. St. Vincent Trade Deficit, 1955–1993

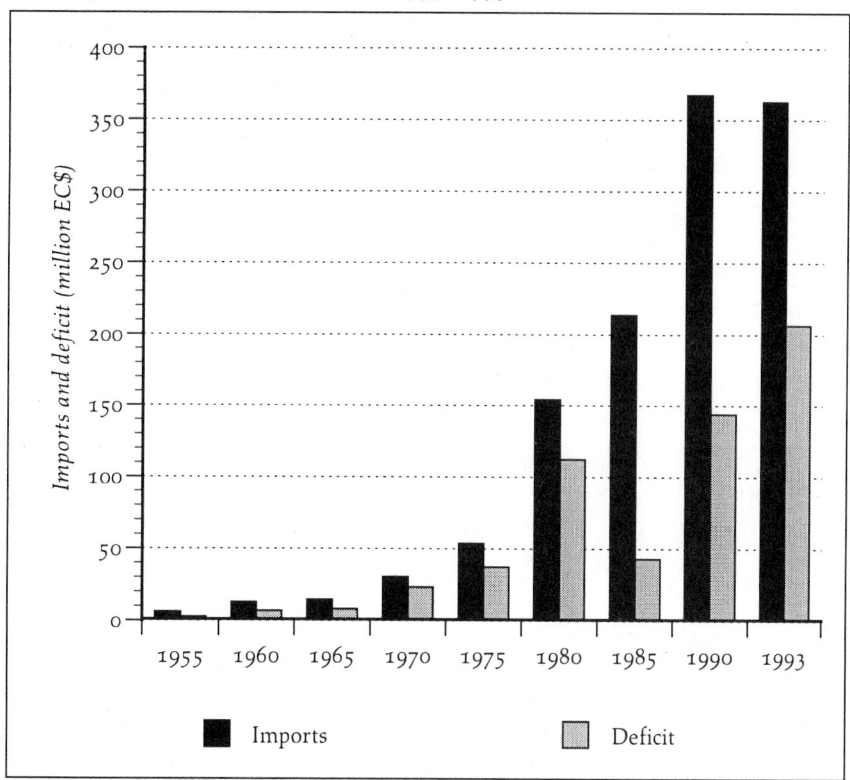

Source: SVG Statistical Unit digests of statistics.

and monuments to such assistance can be found throughout Kingstown—the warehouse of the St. Vincent Marketing Corporation, a statutory corporation that is a major importer, wholesaler, and retailer of food; the new Kingstown fish market; and the modern, five-story Finance Office Complex, which is the heart of the government and the island's tallest building.

In much of the Caribbean, development strategies focus on boosting tourism and manufacturing. Mass tourism has never taken hold on St. Vincent, despite its rugged beauty, as it has in nearby Barbados, in part because of the inability of the island's airport to handle large jet traffic and the lack of extensive, white beaches. Rather, the destination for most foreign tourists are the beautiful, small islands and turquoise waters of the Grenadines that yachts and cruise ships frequent. Nonetheless, tourism is beginning to grow moderately, with the number of visitors increasing 37 percent from 1990 to 1996, to 216,000 (SVG Statistical Unit 1995: 63; Fae Nurse, pers. comm., 26 May 1997). In contrast, the manufacturing sector has grown slowly and has been somewhat stagnant over the last several years. Most manufacturing activity, which centers mainly on light industries—electronics, garments, food processing, and packaging—takes place at the island's sole industrial estate at Camden Park, west of Kingstown. Certainly, the much-renowned "new international division of labor" (see Fröbel, Heinrichs, and Kreye 1980), which has brought branch plants to many developing countries, has had little impact on St. Vincent.

Since the establishment of the colony of St. Vincent in the eighteenth century, agriculture has been the mainstay of the economy and bears the burden of providing the exports that are so crucial to this open economy. Agriculture's importance is evident throughout Kingstown. In the bustling food market, especially vibrant on Fridays and Saturdays, several hundred Vincentian women sell a wide variety of fresh, succulent fruits, vegetables, and "ground provisions" (root crops) grown throughout St. Vincent. Similarly, near the port, piles of sacks containing Vincentian ground provisions await shipment to Trinidad. And on Mondays and Tuesdays long lines of trucks clog the roadways as they carry their cargoes of cardboard boxes packed with green bananas destined ultimately for the U.K. market.

In 1993, agriculture's share of the gross domestic product (GDP) was an uncharacteristically low 14.3 percent, reflecting the severe downturn in the banana industry as it struggled to cope with the competitive pressures stemming from the European Union's new Single European Market (SVG Statistical Unit 1995: 67). In the previous five years, when the banana industry was more buoyant, agriculture was the largest contributor to GDP, fluctuating between 16.6 and 19.0 percent (ibid.).

Bananas dominate the agricultural sector, making up approximately 80 percent of the total value of all crop exports in 1992 (SVG Statistical Unit 1995: 61). Although data on the area under bananas are somewhat questionable, official SVBGA statistics (SVBGA files) give a rough approximation of the growth in the area under the crop in the 1980s and early 1990s (Table 3.1). With the severe downturn in the industry starting in late 1992, the area under bananas has fallen since that time, though accurate official estimates are unavailable. And just as the banana is the most important crop on the island, so too is the industry the major employer. Association officials (SVBGA 1994: 3) estimate that in the late 1980s and early 1990s, the banana industry directly or indirectly employed 54 percent of the nation's workforce.

Bananas are certainly not the only important crop.[2] The cultivation of ground provisions, most of which are destined for export to Trinidad, is a traditional endeavor for much of the peasantry. However, crops considered "nontraditional agricultural exports," which play an increasingly significant role in many de-

TABLE 3.1. Area under Bananas, St. Vincent and the Grenadines, 1981–1992

Date	Area under Bananas (acres)
1981	6,000
1982	6,200
1983	6,200
1984	6,438
1985	7,226
1986	7,226
1987	9,770
1988	10,000
1989	10,000
1990	12,000
1991	12,618
1992	12,618

Source: SVBGA files.

veloping countries focused on export substitution (see Thrupp 1994), are of limited importance in St. Vincent.

The Political-Economic and Environmental Contexts of Agriculture

Inequality is a fundamental, long-standing feature of Vincentian life. People are clearly conscious of this reality, as revealed in one Vincentian's comments concerning a case before the courts: "In St. Vincent, if you got money, you can't be wrong."

Historically, inequality was nowhere more evident than in the distribution of land, though the extent of maldistribution has decreased considerably over time. The most recent agricultural census in 1985–86 reveals that for St. Vincent and the Grenadines, 5,862 of the 6,799 holdings were less than 5 acres, and an additional 672 ranged in size from 5 to 9.9 acres (Table 3.2). All agricultural censuses in St. Vincent are open to question concerning their accuracy, and the 1985–86 census is no exception, especially because of the limited involvement of the Ministry of Agriculture in data collection. In particular, the accuracy of data concerning the number and area of holdings less than five acres is especially questionable and likely underestimated. Nonetheless, the numerical dominance of the peasantry in Vincentian agriculture today is clearly evident.

In contrast, census data revealed that the area in holdings over one hundred acres, which are likely to be most accurate, was 13,395 acres. Results from the

TABLE 3.2. 1985/86 Agricultural Census of St. Vincent and the Grenadines

Size of Holding (acres)	Number of Holdings with Land	Area (acres)
Less than 5	5,862	7,586
5–9.9	672	4,058
10–49.9	215	3,577
50–99.9	16	1,078
100–499.9	25	4,221
500 or more	9	9,174
Total	6,799	29,694[a]

Source: SVG Ministry of Agriculture, Industry, and Labour 1989: 177.
[a]Of this total, only 1,819 acres were in the Grenadines.

1946 and 1972 censuses indicated that 22,114 and 18,407 acres, respectively, were in such holdings (SVG 1978: 23, 38), reflecting the gradual decline of the estate sector.[3] Such large-scale holdings included both private and government estates, as the state has long been involved in purchasing private properties for eventual redistribution in land reform programs. The pace of government acquisitions of estates for such purposes accelerated in the 1980s, by which time it was clearly the largest landowner. And starting in the late 1980s, the government, with the help of external funding, began redistributing its ten estates, consisting of over eighty-five hundred acres, in small plots to expand the number of peasant holdings and to increase exports of agricultural products.

The role of the state in Vincentian agriculture is certainly not confined to land reform. Its road-building efforts, besides serving to increase government popularity and gain votes, have been designed to encourage export agriculture, especially banana production. But government outlays in support of local food production have always been limited, a pattern that has plagued St. Vincent and other countries of the English-speaking Caribbean (Long 1982; Axline 1986). Indeed, such outlays accounted for only 3 percent of the government's recurrent budget (SVG Statistical Unit 1995: 39), leading to an inadequately funded agricultural extension service.[4] Although many peasants have obtained land through the assistance of the Vincentian government in various land reform programs, whatever success they have had in domestic food production and marketing must be attributed primarily to the peasants themselves and not to continuing government support.

Also negatively affecting agriculture is the natural environment. Rainfall is not uniformly distributed seasonally or spatially. A dry season extends from January to early May, with the least rainfall occurring between February and

April. Drought is a periodic problem during this time, accentuated by the porous nature of Vincentian soils and steep slopes. Most rainfall occurs between June and November (Birdsey et al. 1986). Precipitation also varies according to altitude, with the driest areas along the coast receiving only 60 to 70 inches per year, while higher elevations inland, which benefit from orographic rainfall, receive as much as 150 inches per year or more, especially above 1,500 feet.

Another serious constraint is the predominance of rugged terrain throughout much of this volcanic island, especially in the densely wooded interior (see Map 3). A central, mountainous spine runs roughly north to south, punctuated by several dramatic peaks, the highest being La Soufrière volcano in the north at slightly over four thousand feet. The spine has numerous lateral ridges descending to the windward (eastern) and leeward (western) sides, with the latter being more mountainous. The distribution of areas under various slopes is particularly illustrative (Table 3.3). Thus, over 80 percent of the land surface has slopes of twenty degrees or greater, hardly an ideal setting for agriculture. Indeed, application of traditional land-classification schemes dealing with suitability for agriculture has limited relevance in understanding Vincentian land-use patterns, because what some authorities consider unsuitable for agriculture is readily cultivated by Vincentian farmers. Some farms have slopes over forty degrees! Given such rugged terrain and a land distribution pattern that has relegated much of the peasantry to the most marginal land, soil erosion has always been a serious threat.

The difficult terrain also constrains the transportation system. Given the steep slopes of the interior, no road can bisect the center of the island from east to west. Rather, all major vehicular travel is along the coast, with spurs directed toward the inland valleys on each side. Most roads, especially as one leaves the main coastal roads, are narrow and winding and in constant need of repair. The poor state of roads is a frequent complaint of banana farmers, whose tender, fragile fruit can easily be bruised when carried by bouncing trucks trying to avoid the plethora of potholes as they speed toward Kingstown.

Also relevant to agriculture is population pressure on resources. The most recent population census, taken in 1991 (SVG Statistical Office 1993), indicated that the island of St. Vincent was occupied by 98,132 people, which yields a population density of 733 per square mile. But such an already high figure is misleading because so much of the land is not suitable for settlement or agriculture. A generous definition of the word would designate roughly forty thousand acres on St. Vincent as "cultivable," resulting in a much higher "effective" population density of 1,570 per square mile in 1991 (United Nations Development Program Physical Planning Project 1976a: 3). Fortunately, the recent rate

MAP 3. Topography of St. Vincent

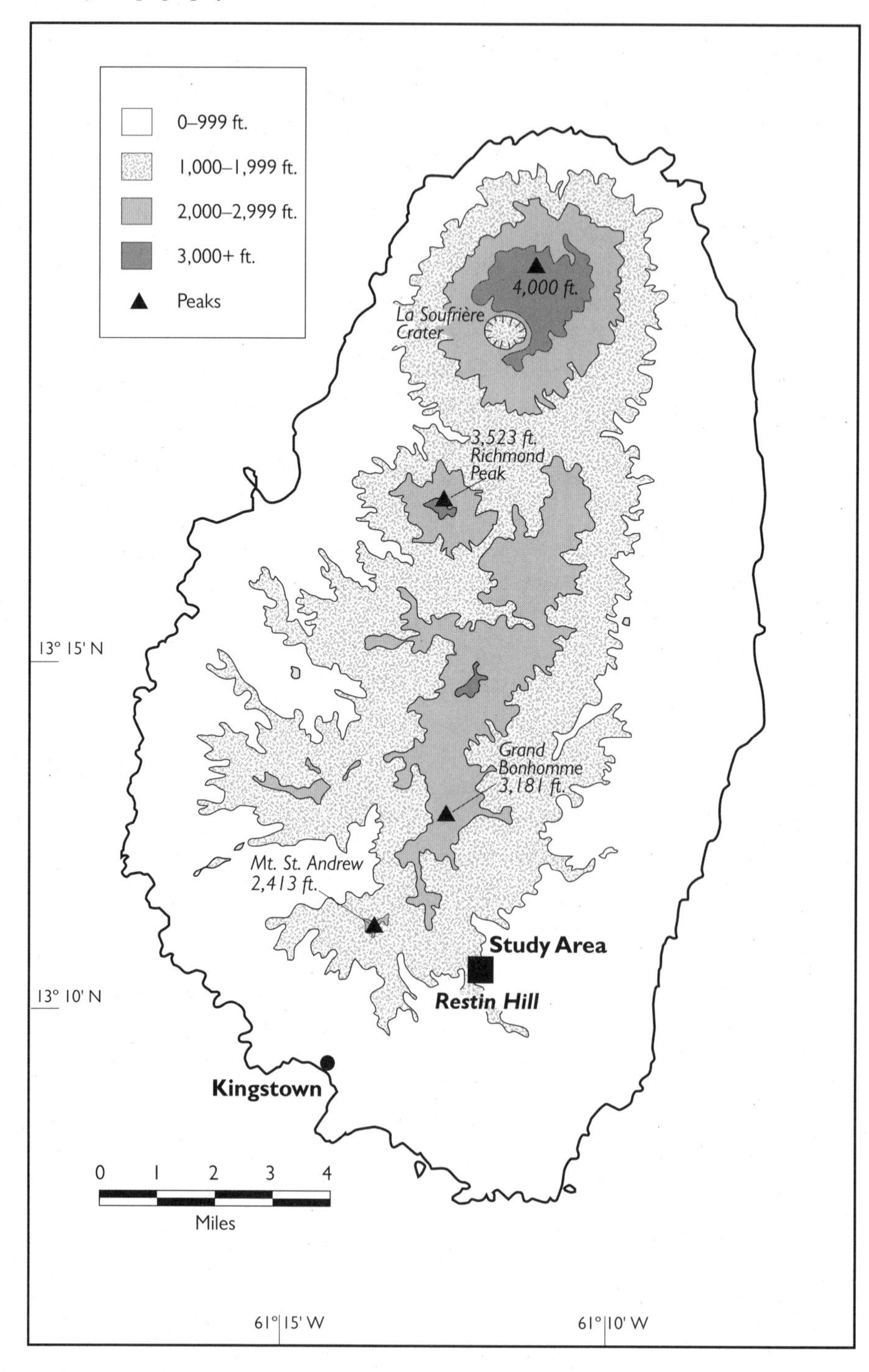

TABLE 3.3. Distribution of Land Area according to Slope Classes on St. Vincent

Slope (degrees)	Area (acres)	Percentage of Total
0–2	3,298	3.87
2–5	1,828	2.15
5–10	5,451	6.40
10–20	5,930	6.96
20–30	23,355	27.43
>30	45,268	53.19
Total	85,130	100.00

Source: Ahmad 1987: 126.

of population growth has been comparatively slow, only 0.77 percent per year from 1980 to 1991 (which has fallen from the previous rate of 1.17 percent per year during the 1970–80 period) (SVG Statistical Office 1993: 4). Emigration has been a particularly important safety valve; indeed, visits to any Vincentian community will yield numerous inhabitants with relatives in the United States, Canada, Britain, Trinidad, and Barbados.

Nevertheless, in spite of the slow rate of population growth on the island, limited land availability is a constraint in agriculture. In most rural areas, a considerable number of villagers do not have land for farming—a pattern widespread in the Caribbean (Mintz 1974). In the Vincentian context, no official figures are available, but it is possible to estimate that the nonfarming village population is approximately 15 to 20 percent of rural inhabitants. Moreover, given the unequal distribution of land, many who do engage in agriculture have insufficient holdings to be self-supporting. Thus, a sizable percentage of the rural population requires off-farm employment. But unemployment continues to be a pressing problem. Reports in the 1980s indicated unemployment levels of between 23.5 and 40 percent, the amounts varying according to the definitions of unemployment used, while the most recent census in 1991 recorded a rate of 19.8 percent (SVG Statistical Office 1993: 44). Such problems are reflected in the GDP per capita for 1993, which was only EC$4,971, low for the region (SVG Statistical Unit 1995: 5, 66). And recently depressed conditions in the pivotal banana industry in the 1990s have certainly intensified the unemployment situation.

The structure of farming and the inequalities in agriculture, which are so fundamental for comprehension of the political ecology of banana contract farming, are not recent creations. Understanding their origins as well as their environmental implications requires consideration of the evolution of agriculture and the rise of the peasantry.

Sugar and the Vincentian Colonial Economy

The sugar cane–based plantation system—the hallmark of colonial economies in the British Caribbean—came comparatively late to St. Vincent. When planters first exported sugar from the nearby British colony of Barbados in the 1640s (Rubenstein 1987: 28), Carib Indians still controlled St. Vincent, having resisted attempts at European settlement. The French from Martinique were the first to gain a foothold in 1719, but British forces eventually captured the island in 1762 and obtained formal control in the Treaty of Paris the following year. Except for a brief interlude under French control from 1779 to 1783, the island remained under British authority until independence in 1979.

The new British colonists soon turned their focus to sugar cane—a crop that Trouillot (1988: 55) calls "that almost inevitable aspect of island history in the Caribbean." By 1771, planters from North America, Barbados, and Antigua obtained over twenty thousand acres on St. Vincent (Wright 1929; David Watts 1987: 294). Unfortunately for the Caribs, their remaining lands were also prime sugar cane areas, which enticed the new colonists to encroach on their territory. The Caribs, in turn, resisted, which led to major conflicts with the settlers in the 1770s and 1790s. The British finally defeated the Caribs in 1796, expelling the large majority to a small island off the coast of Honduras (Gonzalez 1988).

Sugar cane cultivation soon dominated the Vincentian landscape, reaching its zenith on the island in 1828, when colonists exported 14,403 tons (Spinelli 1973: 67). As elsewhere in the Caribbean, it was based increasingly on large-scale enterprises. The expansion of production, which required considerable capital for equipment to process sugar, molasses, and rum in estate factories, led to land consolidation and the demise of small-scale cultivators (Rubenstein 1987: 28; see also Howard and Howard 1983: 77). For example, in the late 1700s, there were 188 estates on St. Vincent, with an average area of 93 acres (John 1974: 91), but by the late 1820s, only 96 remained, with a much higher mean holding of 369 acres (Shephard 1971: vi–xxvi).

Estate sugar production also required a substantial labor force, a problem "solved" by the mass importation of slaves from Africa, who made up an increasingly large percentage of the island's population over time. In 1764, just before commercial exports of sugar, St. Vincent had 7,414 slaves. By 1787, that number had grown to 11,853, and by 1812, to 24,920 (Shephard 1971: iv). A rigid, pyramidal social hierarchy associated with the estate system developed, with the small, white planter class at the top and the slaves at the bottom (see Carmichael 1833; Smith 1965; Rubenstein 1987).

St. Vincent entered the sugar industry past the time of greatest prosperity for British West Indian sugar estates. Indeed, the nineteenth century was a period of great difficulty for sugar producers in the region. In 1807, only forty-one years after St. Vincent first began exporting sugar, the British government abolished the slave trade, which increased the estates' costs of obtaining and maintaining slaves. Soon afterward, in 1812, the Soufrière volcano, which dominates the landscape at the northern end of the island, erupted and devastated the sugar industry.

The fortunes of king sugar continued to decline with the abolition of slavery in 1834. After an initial four-year apprenticeship period ending in 1838, the former slaves were no longer forced to work on the estates, which immediately created fears among planters about possible labor shortages. Harboring bitter memories from slavery and uninspired by the low wages offered, many former slaves preferred not to continue working on the estates. Indeed, the resident labor force on estates declined from 10,667 in 1838 to 8,659 in 1846 (Fraser 1980: 11), as many moved to newly established "free villages." They preferred either to obtain land—by purchase or rental—to establish an existence independent of the estates or to emigrate to Trinidad, where wages were higher (Marshall 1965; Mintz 1985). But their options were, in reality, quite limited. Planters preferred not to sell land to their former slaves, wanting them instead to remain dependent on the estates for their livelihood. Furthermore, leaving residences on the estates did not necessarily mean that the former slaves were no longer dependent on wage labor, as many who left continued to work for planters, at least part-time (Hall 1978). Thus, although labor relations were often strained, most former slaves had to engage in wage labor for survival.

Problems for the sugar estates intensified. Extensive absentee ownership led to poor management, as owners resident in the United Kingdom left their properties in the hands of attorneys who had little vested interest in the efficiency or long-term profitability of the enterprises (Davy 1854: 184). Particularly ominous was the gradual decline in the system of preferential treatment for British West Indian sugar, in which the product from the British West Indies enjoyed lower import duties than imports from other sources (Curtin 1954). This preferential system was fundamental to profitability for British West Indian producers. But the free trade movement was gaining strength in the United Kingdom with the expansion of industrialization, and as sugar was becoming an increasingly important commodity in mass consumption, pressure built for the removal of the preferential import structure that was keeping the cost of British sugar artificially high. Problems accelerated with the passage of the dreaded Sugar Duty Act of 1846, which led to the equalization of all duties on sugar of the same quality by 1854. Worse yet, in 1874 Britain abolished all duties on sugar—irrespective of source. Lower sugar prices and mounting financial problems were the ultimate result for the increasingly desperate Vincentian planters.

As the century drew to a close, the West Indian sugar industry deteriorated further. Competition in the British sugar market intensified as European sugar beet producers, benefiting from more favorable export subsidies from their respective governments in 1884, began shipping larger amounts of sugar to the United Kingdom, causing prices to drop precipitously. Thus began an economic depression in the British West Indies that would last for the rest of the century (Richardson 1989).

The embattled Vincentian industry was close to collapse in the 1890s. In fact, by 1892, arrowroot (*Maranta arundinacea*) replaced sugar as the island's main export. In the period 1895–99, sugar production was only 22 percent of the output at the time of emancipation (David Watts 1987: 496). Environmental disasters further crippled the limping industry. A major hurricane struck in 1886, and an especially devastating one occurred in 1898, soon followed in 1902 by another deadly eruption of the Soufrière volcano, which caused "appalling damage" (Duncan 1970: 48; see also Richardson 1989).

Responding to the depressed state of the region, the British government sent a Royal Commission (West India Royal Commission 1897) to investigate conditions in its colonies and make recommendations. It viewed the prospects of the Vincentian sugar industry in a particularly dim light (ibid.: 47): "The sugar industry has been in a decaying condition for years, and is now on the verge of extinction. No improvements have been introduced in the manufacture of sugar." Consequently, it argued that the number of small-scale proprietors should be increased significantly to alleviate the extensive poverty on the island and help diversify the economy. Although the Commission felt government-sponsored peasant land settlement schemes were appropriate for the region generally, it especially emphasized their need in the Vincentian case. At the time, most of the land remained in the hands of the estates. Indeed, one British merchant firm, D. K. Porter and Company, controlled twenty-one estates containing 11,826 acres (ibid.: 119). But in 1896 only approximately 25 percent of the arable land on the island was being cultivated. Thus, the long-suppressed peasantry would come to play an increasingly significant role in the Vincentian economy, given an initial boost from the policies suggested by the Royal Commission report.

The Origins of the Vincentian Peasantry

Mintz (1961, 1985, 1989; Mintz and Hall 1960) has made a fundamental contribution to the Caribbean literature in his analyses of the historical basis of the contemporary peasantry. In particular, he has asserted that slave involvement in producing food both for home consumption and for sale and participation in marketing formed the basis of a "proto-peasant" adaptation that was the precedent for subsequent peasant activity in the postemancipation era. Such antecedents clearly distinguish the origins of Caribbean peasantries from the situation in most other regions of the world, where previously independent tribal groups became peasants as they lost their autonomy to new colonial states and the pressures of new commodity markets. Tomich's (1990: 260) comments concerning the history of nearby French-controlled Martinique are relevant to the Vincentian case: "Thus, it was not a traditional peasantry attacked from the outside by commodity production, the market economy, and the colonial state; rather, it was formed from within the processes of the historical development of slavery and the plantation system." Caribbean peasantries emerged from slaves who adapted to and resisted estate domination, with the peasantry continuing that tradition of struggle against estate dominance in its formative years.

Certainly, differences between peasants and slaves are evident in their social relations of production. Marshall (1968) notes that, unlike peasants, slaves controlled neither their land nor their labor. Where and when slaves could cultivate were determined by estate owners and legislative decrees. But control by capital over labor in slavery was certainly not unproblematic. Estate owner complaints about lackluster work performance by slaves were widespread; Carmichael (1833: 96), wife of an estate owner, lamented, "Employment is their abhorrence—idleness their delight." Observers of the time also contrasted the zeal with which slaves cultivated their own gardens with their half-hearted efforts in the sugar cane fields (Rubenstein 1987). Moreover, although when slaves could work in their own gardens was determined by estate owners and legislative decree, the actual organization of labor within their plots, the range of crops grown, and the amounts of produce sold were all decisions that the slaves made themselves.[5]

Food production was a key component of the proto-peasant adaptation, but slave involvement in such activity was not uniform throughout the British West Indies. Stephen (1830: 261) distinguished between "foreign-fed" and "home-fed colonies." The former were those colonies, primarily in the Leewards, that depended mostly on food imports to feed their slaves; such colonies had extensive areas of prime agricultural land that estate owners preferred to allocate to sugar cane cultivation instead of food crops. The latter, which included the Windward Islands and Jamaica, were those that relied primarily on slaves to produce food for themselves and the rest of the islands' inhabitants; these colonies had a larger percentage of rugged, mountainous land not suitable for sugar cane cultivation compared with the "foreign-fed" colonies. This pattern of allocating the best land for large-scale, estate export agriculture and relegating small-scale, slave production of domestic food crops to more marginal, rugged land in the interior of the islands left a lasting spatial imprint on the landscapes of these colonies, including St. Vincent.

In the case of St. Vincent, slaves had access to three different types of land for food production (Marshall 1991). Yard gardens were small pieces of land surrounding slave huts where they cultivated a variety of crops, such as cassava, beans, tannia, peas, pigeon peas, and cabbage and raised livestock as well (Carmichael 1833: 136). As slave huts were typically located on land not suitable for sugar cane cultivation, the productivity of such land was likely quite limited (Marshall 1991). Slaves also were granted access to small "yam pieces" in areas where sugar cane had been harvested; cultivation of such plots with yam and other crops and application of animal manure helped prepare the land for subsequent replanting with sugar cane (Colthurst 1977: 171; Cameron 1967: 12).[6] Their largest plots, commonly known as "provision grounds," were located in rugged lands in interior areas not suitable for sugar cane, poor in quality, and often a considerable distance from slave huts. Here slaves grew a wide range of crops, especially "ground provisions" (sweet potatoes, yam, cassava, eddoes, and tannia) as well as plantains, bananas, cabbage, corn, beans, carrots, turnips, tomatoes, pumpkins, pigeon peas, and breadfruit trees (Carmichael 1833: 162–73). They also grew arrowroot, a crop originally cultivated by the Caribs, for use as an infant feeding supplement, for medicinal purposes, and for sale. Indicating the links between the proto-peasant adaptation and subsequent peasant activity, arrowroot would later become the main export cash crop of peasants in the postemancipation period (Handler 1971). Estate owners allowed their slaves to work in their provision grounds each Sunday, the only regular day off each week, as well as on alternate Saturdays or a half-day each Saturday when the cane harvest was not in operation (Carmichael 1833: 174; Marshall 1991: 54; Higman 1984: 182; Cameron 1967: 12).[7]

For estate owners, the granting of access to provision grounds was not an act of benevolence but one of self-interest; it lowered their costs of maintaining their slaves (Mintz and Hall 1960). The only food item regularly imported for feeding slaves was salted codfish (Rubenstein 1987: 35).

At the same time, food production also had distinct advantages for slaves. First, it added variety to their diets compared with that available from imported foods in the "foreign-fed" colonies. Second, it enabled them—at least temporarily—to escape the supervision of estate owners, as they were allowed to determine their own cultivation patterns in their provision grounds. Moreover, slaves were allowed to sell "surplus" crops and livestock at the market in Kingstown each Sunday, giving them a source of income to buy items to supplement their diets and obtain simple consumer goods, such as candles, clothes, jewelry, soap, and tobacco (Carmichael 1833: 138, 178; Marshall 1991: 58). The importance of slaves in the internal marketing system is made clear in Marshall's (1991: 56) excellent discussion of provision grounds in the Windward Islands: "By 1790 planters pointed to the slaves' virtual monopoly of the internal markets for locally produced food, firewood and charcoal, and fodder. Urban dwellers purchased much of their food from slaves and the planters themselves depended on slaves for the greater part of their supply of poultry and fresh meat."

Growth of the Peasantry

The growth of the Vincentian peasantry, as elsewhere in the Caribbean, has been a history of struggle against estate domination, a theme emphasized and aptly captured by Mintz (1985: 131), who writes, "Like blades of grass pushing up between the bricks, the peasants of the Caribbean have been embattled since their beginnings." Consequently, the emerging peasantry exhibited spatial similarities to their counterparts during slavery—they grew "in the crevices" (ibid.), in marginal areas of least use to the estates.

With emancipation, the primary concern of estate owners was their ability to maintain control over labor. Most refused to sell plots to the former slaves, instead adopting alternative relations with their workers; a common method was to pay workers partly in nonwage forms of compensation, including rent-free accommodations, rent-free provision grounds, the right to produce charcoal from estate grounds, and allowances of sugar and rum (Davy 1854: 185). Such forms of compensation can be viewed as a mechanism of labor control because continued access to land was contingent upon the provision of labor to the estates by former slaves (Marshall 1965). Indeed, planter opposition to the growth of the peasantry in the British Caribbean was particularly successful in St. Vincent (Momsen 1987a).

Nevertheless, the "ex-slaves' land hunger was enormous and evident" (Marshall 1968: 255). By 1845, former slaves had purchased at least 158 holdings ranging in size from eleven acres to less than one acre (Davy 1854: 182). The lands purchased, however, were generally poor in quality, as the estates sold off the least desirable parts of their properties first. Squatting on rugged Crown land, which enclosed the interior parts of the island, became another means of gaining access to land (Beard 1947: 299). The former slaves also rented land when made available by the estates. From such properties, they produced crops and raised livestock for subsistence and sale in the local markets—continuing the tradition established during the era of slavery. While estates maintained their focus on sugar cane production, the new peasants dominated the export

TABLE 3.4. Distribution of Land on St. Vincent, 1895

Size of Holding (acres)	Number	Area (acres)[a]
Over 50	134	42,000
20 to 50	34	1,060
Less than 20	351	1,360

Source: West India Royal Commission 1897: 119.
[a]Does not include the Grenadines. Based on holdings of one acre or more. Did not include approximately five hundred additional holdings of Crown land, ranging in area from five to twenty acres, which had not been fully purchased by the time of the survey.

trade in arrowroot, which became the island's second most important export crop in the second half of the nineteenth century.

Administrative officials in St. Vincent were sympathetic to peasants' and workers' growing demands for land, in spite of the opposition of the planters (Fraser 1980). The government thus began selling small, largely five-acre lots of Crown land in the early 1890s.[8] By 1905 it had sold 318 plots totaling 2,438 acres and had also started renting small plots of Crown land (ibid.: 111). The sale of Crown land, in reality, was only a minimal gesture by the state toward expanding the peasantry. Much of the land sold was mountainous, with poor soils, and in remote locations, making access to markets difficult. What made conditions particularly difficult for peasants wanting to purchase Crown lands was the changing cropping patterns on estates. As sugar prices continued to plummet, estate planters shifted to arrowroot production, the export crop originally the province of peasants. As a result, the arrowroot market became glutted and prices dropped, reducing peasant incomes and their ability to buy land (Fraser 1986: 31).

Observers in the 1890s frequently contrasted the dismal economic state in St. Vincent, which still had few peasants and retained a high degree of land concentration, with the more prosperous situation in nearby Grenada, which already had many more peasants (Richardson 1989: 114). Table 3.4 presents a list of the distribution of holdings on St. Vincent in 1895 as provided in the West India Royal Commission report.[9]

Indeed, the Commission (West India Royal Commission 1897: 48) complained about the reluctance of the estates, even those undercultivated, to sell small lots or make them available for sale at reasonable prices. The marked inequality and underuse of land thus inspired the Royal Commission to recommend a land settlement scheme to expand the peasantry. With funding from the British government, the Vincentian administration began buying land in 1899 from estate owners willing to sell and compulsorily purchasing estates that were essentially defunct. By 1915 it had acquired 7,527 acres for land settlement schemes (John 1974: 118), distributing parcels mostly in five-acre blocks (Wright 1929).[10]

The land settlement scheme was plagued by numerous problems (John 1974; Fraser 1980; Momsen 1987a; Richardson 1997). Estate owners willing to sell to the scheme generally disposed of their most unproductive land. Similarly, many of the estates that were already defunct had poor-quality land or exhausted soils. Nonetheless, the land settlement program did help to expand the peasantry significantly (Fraser 1986). But the pattern of purchasing and distributing primarily poor-quality land contributed to the tradition established in the era of slavery—relegating small-scale food production to more marginal areas.

Peasant Agriculture in the Twentieth Century

With the help of the Land Settlement Scheme and given the problems in the sugar industry in the Vincentian economy, the first half of the twentieth century was characterized by the growth of the peasantry and the gradual decline of the estate sector. During this period, two export crops dominated the Vincentian economy—arrowroot and cotton. Peasant involvement in both was considerable.

Arrowroot was first exported from St. Vincent as early as the 1830s (Wright 1929: 254). The crop was particularly suited to peasant agriculture because it could be grown in a wide range of environmental conditions and thrived on the sloping land characteristic of peasant holdings, but production did have one major drawback—the task of harvesting the cylindrically shaped roots was especially backbreaking work. Peasants played a proportionally more important role in the Sea Island cotton industry, which began in 1903.[11] In comparison with arrowroot, the crop, which produces a premium, superfine cotton, was more restricted in its environmental requirements, being grown mainly near drier coastal areas. Cotton was noteworthy for the environmental problems associated with its cultivation. The crop tended to exhaust Vincentian soils and, more important, contributed significantly to soil erosion (McConnie n.d.: 2–3). In relation to both export crops, peasants sometimes intercropped—arrowroot occasionally with cassava and yam, and cotton more often with corn and cassava and, to a lesser extent, sweet potatoes (Wright 1929; Shephard 1945).

St. Vincent became the world's leading exporter of arrowroot and Sea Island cotton (Wright 1929). But the market for both was never very large. Output of arrowroot and cotton fluctuated considerably during the first half of the 1900s in response to changing market conditions and demand for alternative crops (see Spinelli 1973). For example, arrowroot production fell during both world wars as demand for cotton and food crops increased, and the cotton industry suffered particularly during the Great Depression (Walker 1937). Both eventually declined in importance to the Vincentian economy with the advent of less expensive varieties and sources from other countries and the development of substitutes, which curtailed interest in the crops—with the cotton industry collapsing in the 1960s and production of arrowroot, the island's main export for most of the first half of the 1900s, withering substantially shortly thereafter.

Neither cotton nor arrowroot brought prosperity to the peasantry (Engledow 1945). Peasant holdings were on land too small and poor in quality and cash crop prices were too low to yield substantive incomes. For workers on arrowroot and cotton estates, the crops, which had highly seasonal labor requirements, failed to provide year-round remunerative employment, as indicated by the 1938 Agricultural Department (SVG Agricultural Department 1940: 15) annual report: "The general labor position is therefore that the supply of agricultural labor is much in excess of demand, and very few persons have regular employment. The great majority of the labouring class in the Colony probably have less than six months' employment in the year, and on the Leeward coast the majority of those who are employed at all do not obtain more than three or four months' employment in the year." The depressed state of the agricultural sector was reflected in Walker's (1937: 222) observation that "[v]isitors to the island at the present day cannot fail to be impressed by the abject poverty of a great part of the negro population outside the capital town of Kingstown." Indeed, such difficult domestic conditions in the first half of the twentieth century spurred continuing emigration in search of work to a variety of destinations, including Trinidad, Cuba, the Netherlands Antilles, Venezuela, Panama, and Brazil (Fraser 1986: 56).

Nonetheless, the peasantry had grown considerably since the end of the nineteenth century. Browne (1939: 154) reported that by 1937, properties under ten acres contained a total of 11,382 acres, a vast increase since the late 1890s. Shephard (1945: 14) concluded that St. Vincent likely contained as many peasant proprietors as all the other Leeward and Windward Islands together, excluding Grenada. But despite the considerable increase in the area controlled by the peasants, large estates still held the most land, with estates above fifty acres accounting for 37,556 acres (Browne 1939: 154). The growth of the peasantry in the first half of the 1900s was not a function of an increasingly prosperous peasantry accumulating land. Rather, it reflected state involvement through land settlement and the depressed conditions in the estate sector, which led to further sales of small plots.

One of the characteristics of peasant agriculture in the first half of the twentieth century that constrained productivity was limited technological innovation. Although estates started to use synthetic fertilizers, particularly sulfate of ammonia, during this period, peasant adoption of the practice was limited. For example, a survey of farmers on three Vincentian land settlements revealed that the percentage of those using synthetic fertilizers ranged from only 2 to 10 percent (Shephard 1945: 169), with both low incomes and lack of access to credit accounting for the limited rates of adoption. Similarly, pesticide use was minimal compared with practices on the estates. Some estates at the time used such agrochemicals as bordeaux mixture (a combination of copper sulfate and lime) and lead arsenate to control pests in cotton during the 1930s and 1940s. And although estates employed more agrochemical inputs than did peasants, they were certainly not in the forefront of technological innovation themselves. Concerning the level of technological innovation, Jolly (1947: xvii), in an analysis of the Vincentian economy, mused that "[i]t would be no exaggeration to say that the bulk of agricultural production is carried on at no higher technical standard than cereal growing under the old Manorial System of England which became obsolescent six hundred years ago."

An indication of the malaise in peasant agriculture was the relatively large percentage of uncultivated land, a surprising condition given the high population density on the island. Shephard (1945: 154) reported that peasants left an average of 31, 49, and 57 percent of their holdings uncultivated in three Vincentian land settlements he surveyed. Commenting on similar problems on land settlement schemes elsewhere in the British Caribbean, he concluded that "there is clearly something radically wrong when 50%, 60% and even more than 70% of the land remains uncultivated" (ibid.: 45). Indeed, the extent of underuse of agricultural land in the twentieth century has been widely recognized as a general problem in the English-speaking Caribbean (Shephard 1945; Rubenstein 1975, 1987; Brierley 1985, 1988). Certainly, underuse of land in large-scale holdings is a widely acknowledged phenomenon in the development literature (Lappé and Collins 1977), but it is surprising to find extensive underuse of arable land on small holdings.

Observers have advanced a variety of reasons for underuse of land on St. Vincent. Shephard (1945) argued that a considerable area under peasant holdings in land settlements was poor in quality, a condition exacerbated by the fact "that much of the land on the older settlements has been degraded and rendered irredeemable by the peasants themselves" (46). Admittedly, Shephard had very negative views about peasant cultivation practices, but it is also evident that the poor, marginal quality of much peasant land limited yields and encouraged erosion—which reduced the incentive to cultivate. More recently, Rubenstein (1987: 184–91) provided a long list of demographic, ecological, economic, and ideological influences contributing to the underuse of Vincentian land: migration, uneconomic size of holdings, poor quality of land, poor market conditions, and the low status of agricultural work. For the most part, these constraints reflected political-economic conditions—an inequitable distribution of land that had relegated peasants to marginal, small plots that were highly susceptible to erosion; lack of access to credit that, combined with marginal holdings, yielded low incomes; and a negative view of agricultural labor, which reflected the heritage of slavery and the historically poor economic returns in agriculture.

But Shephard (1945) also noted the significance of so-called praedial larceny, the loss of crops due to theft. He reported that "bitter complaints volunteered by many settlers indicate that the losses are serious and that the prevalence of theft is a deterrent to the more extensive planting of food crops" (44). Particularly interesting is his revelation (ibid.) that the problem of theft was more severe with food crops than with export crops, a pattern that has relevance for discussions of bananas and domestic food production today (see Chapter 6):

> Food crops are usually confined to land least suited to the staple cash crop and, as these areas are frequently the most remote from the cultivator's home, they are seldom under continuous observation. Food crops can be consumed by the thief, and even if sold are unlikely to arouse suspicion. But cash crops, such as sugar-cane, seed cotton and arrowroot rhizomes, require processing; the number of buyers is limited, and in most cases they are registered, and required to keep an account of their transactions: thus the thief runs a greater risk of detention.

Indeed, the problem of praedial larceny was widespread in the English-speaking Caribbean (Engledow 1945: 12).

The malaise in peasant agriculture was also reflected in an aging of the agricultural labor force, a pattern characteristic of agriculture elsewhere in the region (United Nations Development Program Physical Planning Project 1976b: 15; Brierley 1988: 15). For many, agriculture became the occupation of last resort. The failure of agriculture to retain the interests of young men and women resulted from the low status of farming (reflecting, in part, the association of agriculture with slavery: see Finkel 1964; Long 1982), poor returns, and uncertain markets—disinterest that intensified as nonfarm employment opportunities became increasingly available in the post–World War II period. Emigration to the United Kingdom in the 1950s and early 1960s was another strategy employed by those giving up unremunerative agriculture (Rubenstein 1983).

As the contemporary banana industry emerged in the 1950s, the island's economy in general and the peasantry in particular were still in a depressed state. Indeed, economic assessments of the island at the time continued to emphasize the impoverished conditions there (see Fentem 1961). A review by a mission from the University of the West Indies (1969: 8–9), in a bit of hyperbole, exclaimed that "most of the people on the island are living in a way which, in terms of material and environmental conditions could scarcely be far removed from the situation as it was under slavery."

But by midcentury the dominance of the estates was also slowly withering (Fraser 1986), a process that accelerated markedly in the 1970s and 1980s. The demise of the private estate sector reflected a variety of pressures. One was poor prices for agricultural commodities, especially coconuts, a major estate crop. Another was contentious problems with labor. Furthermore, selling land and investing in other segments of the economy, especially wholesaling and retailing, have been more profitable—and less risky—than agricultural production. And subdivision of land among heirs has been significant as well. Indeed, Fraser (1986: 27), in his discussion of the rise of the Vincentian peasantry, challenged the applicability of the "plantation economy" model (see Beckford 1972), which emphasized the continuing dominance of the plantation sector over the peasantry and over economic, social, political, and cultural life in general; he asserted that even by the 1950s, the estate sector was "seriously weakened, it was on the defensive, and in retreat." Thus, since the mid-1900s, the significance of peasant farming has increased dramatically and now dominates Vincentian agriculture. In particular, peasants have made up the majority of banana growers since the establishment of the contemporary banana industry.

Certainly, continuities from the past are evident in agriculture today, patterns that affect the practice of contract farming—unequal control over land, the importance of subsistence production and food marketing, and the tradition of intercropping food crops with export crops. Another persisting pattern is the inability of Vincentian export agriculture—like that elsewhere in the Eastern Caribbean—to survive and prosper without some form of market protection, a problem that sugar cane–growing estate owners of the 1800s and banana producers of today have experienced. At the same time, characteristics once associated with Vincentian agriculture—youth disinterest in farming, underuse of land, low levels of technological innovation, limited agrochemical use, seasonal agricultural employment, and relative noninvolvement by capital and the state in regulating the labor process—have all changed to a considerable degree in response to the growth of banana production based on contract farming. Moreover, the banana industry, while never bringing long-lasting prosperity to the peasantry, certainly did improve their living conditions. As one elderly Vincentian farmer reflected, "Since banana come in, it make a little lift." One village intensively involved in the banana industry and impacted by these changes has been Restin Hill.

Four

Life in a Banana-Producing Village

Traveling inland from the windward coast you climb a narrow, winding road. After a fifteen-minute drive from Kingstown, the breathtaking and dramatic, crater-shaped Marriaqua Valley (also known as the Mesopotamia Valley) suddenly appears on the left—one of the island's prime agricultural areas. Steep, rugged hills enclose this verdant valley, culminating in the majestic peak of Grand Bonhomme (3,181 feet). This inland valley of some twenty villages inspired the British army's chief medical officer, John Davy (1854: 191), to pen this description over a century ago:

> Passing to the windward side, and deviating a little from the line of coast, the fine circular valley of Marriaqua may be seen,—a little district within itself, about eight or ten miles in circumference, surrounded by high hills and mountains, the greater part of it laid out in sugar estates, forming a whole of peculiar beauty. This charming valley, perhaps once the enormous crater of a volcano, and afterwards, it may be, the bed of a lake, opens abruptly towards the coast by a remarkable gorge, through which one of the largest streams of the island finds its way to the sea by a succession of pools and rapids, often peculiarly picturesque from the accompaniments of rock and wood.

Reflecting his cultural background, Davy naturally focused on the Marriaqua sugar estates that dominated the valley at the time. But given the rugged nature of much of the terrain and high rainfall in the valley, the area was also a major arena for provision grounds during the era of slavery and of subsequent peasant cultivation in the early postemancipation era. As elsewhere on the island, the dominance of sugar cane in the valley gradually declined. Elderly informants recall that, in the first half of the 1900s, arrowroot and sugar cane were found on both estates and peasant holdings in the valley. The other pillar of the Vincentian export economy in the first half of the twentieth century, cotton, was rather rare, as the valley's high rainfall was not conducive to its growth. Estates remained important in the valley at the turn of the century, when several of roughly two hundred acres in size existed, accompanied by a host of smaller ones. Even as late as the 1950s, a few estates over two hundred acres remained, but the estate sector has declined considerably through sale and subdivision among heirs. By the late 1980s, only a few medium-sized estates of roughly fifty to sixty acres existed.

Restin Hill, nestled in the back of the valley, was established as an free, independent village before the 1861 census, at which time its population was 140.[1] Other peasant settlements sprang up in the valley in the early postemancipation era, with an adjacent village established as early as 1846. In the first half of the twentieth century, Restin Hill itself was surrounded by estates approximately fifty to seventy acres in area, but they too have declined in size. As elsewhere on St. Vincent, many in Restin Hill had to depend on a combination of cultivating their own lands and working on estates for wages. Peasant ownership of land was less prominent then compared with today, with many villagers also renting or sharecropping plots owned by estates.

Restin Hill was traditionally one of the more isolated communities in the valley. The main roads bypassed the village, and as late as the 1960s, there was only a six- to eight-foot-wide, unpaved dirt track leading into Restin Hill. A paved spur now runs into the community, and the winding, bumpy trip to Kingstown by vehicle takes approximately twenty to twenty-five minutes. As members of only four households in the village own vehicles, the rest of the inhabitants of Restin Hill travel to town mainly by paying the EC$3.00 fee for a ride on a usually crowded passenger van.

The Village Scene

Most of the sixty-two houses sheltering the 333 people living in Restin Hill are laid out along or near the paved road into the village.[2] Two types can be found: "board houses," made primarily of locally sawn wood, and "wall houses," made of concrete blocks, both being covered by galvanized roofs. The former, which include 65 percent of the structures in Restin Hill, are generally smaller and owned by those with lower incomes, with some as small as ten feet by fourteen feet. Normally, a small kitchen, measuring roughly fifty-five to seventy-five square feet in floor area, is contained in a separate wooden building nearby. The more durable "wall houses" are growing in number, with the general trend being for people to upgrade from board houses to larger wall houses as their incomes permit, a trend greatly facilitated by the banana boom in the latter part of the 1980s.

Regardless of the number of rooms in a house (which ranged from one to nine), the sitting room has a fairly standard pattern. Prominent is the china cabinet, which proudly displays a family's dishes and glasses. Interior walls are adorned with Christmas cards, photographs, and posters with a variety of religious expressions. And particularly important as the centerpiece for household entertainment is the ever popular television set, found in 37 percent of the houses in 1989. Even some houses that did not have electricity at that time still had televisions, which were operated off of car batteries. Shows featuring Caribbean music and soap operas originating from the United States are especially popular fare. And a few homes have VCRS.

Although poverty exists in Restin Hill, villagers there assert that living conditions have clearly improved over the last twenty to thirty years, primarily because of benefits from the banana industry, just as Barrow (1992) found in her study of a banana-producing village in St. Lucia. As is true for Vincentians in general, the extent of the hardships that people in the village experience is clearly less than the grinding poverty that plagues much of sub-Saharan Africa, Asia, and Latin America.

Houses are not the only important structures. The Caribbean institution of the rum shop is central to village life. Restin Hill had two such establishments in 1989, both owned by men and one of which had been in operation for over twenty-five years. Men patronize such shops to gather together and "cool out" after work, as well as on evenings and weekends to enjoy some rum, beer, and conversation and to play the much-loved games of dominoes and cards.

Rum shops are also critical components of the Vincentian food system; they provide a locally available source of imported foods and supplies. Instead of having to go to town, men, women, and children can buy basic foods at the shops, albeit at somewhat higher prices compared with supermarkets in Kingstown. In addition to alcoholic beverages, shops sell sodas; frozen chicken, turkey, and beef; salt fish; sugar; rice; flour; milk; a variety of canned foods; and a host of consumer goods, from lightbulbs to toilet paper. The larger Restin Hill shop carried an inventory of forty-seven food and nonfood items.

Another crucial dimension of the rum shop is the provision of credit, which is especially important for those with lower incomes. Shop owners generally grant credit in proportion to the incomes of villagers. Yet the horrors associated with credit and indebtedness described for other regions of the world, linked to high rates of interest charged, are absent here, for shop owners do not charge interest on debts. And because no security is pledged for debts, a shop owner's only recourse if someone defaults is termination of additional credit. Indeed, one shop owner in a nearby village, frustrated by the difficulty in collecting from those indebted to him, posted a sign outside his establishment:

In God We Trust
In Man We Bust
No Credit
Thank You

Although drinking of alcoholic beverages in rum shops is a regular feature of social life for men, for the large majority of villagers, alcohol consumption does not interfere with productive activities. Nevertheless, some men do spend considerable sums on drinks for themselves and their friends, as there is always substantial social pressure to buy beer or rum for others. Some men prefer to stay away from the shops, lamenting, "When the rum is in, the wit is out," noting that occasional quarreling and "confusion" can make the scene unpleasant. In particular, spending too much money on alcohol can lead to considerable friction between men and women of the same households.

Restin Hill Households and Gender

Operationalizing the concept of "household" is forever a thorny issue in the literature, and the case of St. Vincent is no exception (see Rubenstein 1987). While the concept is useful for purposes of sampling and has real significance in everyday life, its boundaries are often not precise. Also, as Rubenstein (1987: 292) notes, it is crucial to distinguish between "household" and "family"; children do not always live with their parents, as some reside in the households of other relatives, particularly that of their grandmothers. Restin Hill households are coresidential units that share meals, raise children together, and generally cooperate economically in the form of joint agricultural activities or the sharing of some income for household expenses. Based on this definition, Restin Hill has sixty-four households living in the sixty-two dwellings.

In contrast to the situation found by Rubenstein (1987) in his study of a leeward Vincentian community, the overwhelming majority of members of Restin Hill households involved in farming cooperate in agricultural activities, even though men and women often control different plots of land.[3] Similarly, in nonagricultural households, single adults living with their parents or sisters usually contribute at least some money from their wages to their households, although a few exceptions existed.

Gender of household head is a significant influence on labor patterns and agricultural activities in the Caribbean (Momsen 1987b, 1988, 1993). In Restin Hill, a slight majority of the households, 55 percent, are headed by men. The high percentage of female-headed households, characteristic of the region, reflects a variety of processes: male migration, widowhood, and the instability of relationships. Male-headed households are generally composed of either married or common-law couples and their children, and some include a variety of other relatives.[4] In female-headed households, many heads are involved in "visiting unions" with a male of another household, who contributes money to

TABLE 4.1. Differences between Male-Headed and Female-Headed Households, Sample Households

	Male-Headed	Female-Headed
% of land under cultivation	82	68
% of cultivated land under bananas[a]	80	70
% of cultivated land under local food crops[a]	26	35

Source: Based on analysis of gardens of twenty-four sample households (twelve male-headed and twelve female-headed households); Grossman 1993.

[a]Data for land under bananas and food crops sum to over 100 percent because some land has both bananas and food crops.

the woman on a regular basis, although many such relationships are not of long duration. Such households also sometimes include a brother of the head, and the addition of a parent or grandchildren of the female head is not uncommon.

Although male-headed households are, on average, only marginally larger than female-headed units—having a mean of 5.29 individuals per household compared with 5.14—labor supply for agriculture is less problematic in those that are male-headed because more active adults are usually present to help. Also, female-headed units tend to be headed by adults who are slightly older (48.34 years) compared with their counterparts in male-headed units (46.24 years).

More significant, substantive differences in relation to land use exist. In particular, labor availability affects the extent of cultivation and the emphasis on the types of crops produced, as revealed in data from a sample of twenty-four households in the village (see Table 4.1).[5] Female-headed households tend to cultivate less of their land, have a somewhat smaller percentage under bananas, which are very labor intensive, and a somewhat higher proportion under food crops, which are less labor intensive. Nevertheless, both types of households devote a majority of their land to bananas. The pattern of female-headed households having a greater emphasis on local food crops compared with that in male-headed units is characteristic of the Eastern Caribbean (Momsen 1988).

Although marriage is the professed cultural ideal, only 23 percent of the households contained either a married couple (16 percent) or a widowed individual (8 percent); instead, common-law unions predominated. Traditionally, only those economically well-off married, but that pattern has been changing somewhat recently, with the number of marriages among lower-income house-

TABLE 4.2. Relation between Wage Labor and Landholdings, Restin Hill

Household Head, Wife/ Common-Law Spouse, or Both Regularly in Wage Labor[a]	Number (%) Controlling Two Acres or Less	Number (%) Controlling More Than Two Acres
No	14 (34)	19 (86)
Yes	27 (66)	3 (14)
Total	41	22

Source: Survey of sixty-three of the sixty-four village households. (One household is excluded because reliable data for it could be obtained only on the number of plots it cultivated, not the area it controlled.)
[a]On either full-time or part-time basis

holds increasing. Whether containing a married couple or not, households are not income-pooling social units. Although both men and women contribute jointly to some household expenses, they tend to keep their own incomes separate. In fact, men and women are often secretive concerning their cash resources.

Moreover, designating households as being "headed" by an individual is not meant to convey the impression that household heads dominate decision making entirely and have complete control over household labor. In some cases, the head clearly does dominate in such matters, but the pattern is not uniform. In some situations, men and women in the household jointly determine their agricultural activities, whereas in others, they are more independent in planning their own activities, though they still usually assist each other in certain agricultural endeavors. Similarly, although children often help their parents, they too may carry out certain economic activities independently.

Although Restin Hill is primarily an agricultural community, wage labor is also an important source of income for many households—a pattern characteristic of all rural Vincentian communities, one persisting from the time of emancipation. The major types of employment included wage labor in agriculture, road repair, trade work, and housekeeping (for people in other villages). Generally, those households in which the household head, wife/common-law spouse of the head, or both were engaged regularly in wage labor tended to control less land than others in which they did not. Thus, of the thirty (48 percent) households in which the head, wife/common-law spouse, or both engaged in wage labor on a regular basis, 90 percent controlled two acres of land or less (see Table 4.2). In contrast, in those households in which neither the head nor the wife /common-law spouse of the head engaged in wage labor on a regular basis, only 42 percent controlled two acres or less.

Land Tenure and Land Use

Perhaps the most contentious aspect of village life involves control over land. When I asked one villager about patterns of land tenure, he summed up what he felt was the most fundamental principle in Restin Hill—"fighting." Disputes over boundaries and ownership are numerous, being the main source of conflicts among households, a reflection of the general scarcity of land and of the symbolic and economic significance of its ownership and control. As the value of land escalated in the 1980s and early 1990s, partly in response to increased earnings associated with banana production, conflicts over control of land have been exacerbated.

Even though no large landowners live in the village—the largest household holding was 15.60 acres—considerable inequality in control over land exists (Table 4.3). The data reveal the small size of most holdings, which is characteristic of St. Vincent. Eighty-six percent of the agricultural holdings, which I define as those of one-tenth acre or larger, were five acres or less. Moreover, 19 percent of all Restin Hill households did not have any agricultural holdings.

Not only is land unequally distributed among households in general, but patterns are also affected by gender. Similar to the situation elsewhere in the Caribbean (Momsen 1988), male-headed households control, on average, more agricultural land, have more farm plots, and have the largest holdings (Table 4.4). Moreover, although female-headed households make up 45 percent of all households, they comprise 67 percent of the nonagricultural households, those with less than 0.10 acres.

Compensating somewhat for the imbalance in agricultural holdings in the village is the receipt of remittances from overseas. Women tend to receive such funds both more frequently and in larger amounts than men. For some women, it was their major source of support. Similarly, women in visiting unions also receive regular income from their companions. But such support is still not enough to overcome the disparities of land-based wealth among households in the village.

Villagers hold land under a variety of forms of tenure: ownership, rental of land from either private individuals or the government (Crown land), lease, sharecropping, family land, "generation-to-generation land," and temporary-use rights. In Restin Hill, the most widespread form of tenure in 1989 was ownership (Table 4.5). Ownership, which can be obtained through either purchase or inheritance, is clearly the preferred form of tenure. It is well known in the literature on the Caribbean that the significance of land ownership extends beyond its intrinsic economic value (Lowenthal 1961; Besson and Momsen 1987; Rubenstein 1987). Restin Hill villagers emphasize that ownership sig-

TABLE 4.3. Distribution of Control over Agricultural Land, Restin Hill

Acres of Land Controlled	Number (%) of Households
No agricultural land	12 (19)
0.10–0.50	10 (16)
0.51–1.00	7 (11)
1.01–2.00	12 (19)
2.01–3.00	7 (11)
3.01–4.00	5 (8)
4.01–5.00	3 (5)
5.01–7.50	6 (9)
7.51–10.00	0 (0)
>10.00	1 (2)
Total	63

Source: Survey of sixty-three of the sixty-four village households.

nifies being "independent." In the abstract, "independence" refers to not having to work for someone else, though ownership of just small amounts of land may still necessitate some involvement in wage labor. Nonetheless, ownership increases the degree of independence and clearly confers a strong sense of pride. Given the history of slavery and the drudgery in past years of laboring for estates, ownership of land has great symbolic significance. Parents also point to the pride of being able to pass their land on to their children to help them get a start in life.

People are particularly covetous of their holdings and will fight attempts to encroach on their properties. Boundaries are marked by the ubiquitous woody plant called "white dragon." Even seemingly minor transgressions, such as the leaves of a neighbor's plants or trees hanging over one's property, can lead to conflicts if relations between neighbors are poor.

The area of land under ownership in 1988, 32 percent, is certainly higher than it was in the past, when few could afford the cost of land. While land ownership is more extensive today, the price of land has also escalated dramatically in the 1980s, in response to the increase in banana incomes and electrification of the village in 1984. In the early 1990s, a plot of one-tenth acre along the road in Restin Hill cost between EC$4,000 and $5,000, though that was inexpensive compared with similarly sized plots in other, nearby villages. Prices for agricultural land varied, depending on size, quality, and distance from a road. Compared with the 1970s, land prices appear to have increased between 200 and 500 percent or more.

In discussions of past purchases, villagers fondly remember the sale of cattle or other livestock that helped them make their initial down payments. Today,

TABLE 4.4. Differences in Agricultural Holdings, by Gender, Restin Hill

	Male-Headed Households	Female-Headed Households
Mean agricultural land controlled (acres)[a]	3.01	0.96
Number of agricultural plots per household[b]	2.85	2.10
Largest household holding (acres)[a]	15.60	3.45
Number (%) of nonagricultural households[b]	4 (33)	8 (67)

Source: Survey of sixty-four village households.
[a]N = 63
[b]N = 64

income from banana sales, supported by loans from banks or credit unions, pays for most properties.[6] Banks had been especially willing to provide mortgages to banana growers, as monthly mortgage payments could be deducted directly from growers' accounts at the SVBGA, but the recent downturn in the industry has forced some banks to take a more cautious attitude toward agricultural loans.

As much as land is in demand, very little is for sale. As Rubenstein (1987) found, villagers sell land only when they have left permanently or are in dire financial circumstances. People who want to sell land do not have to advertise to find willing buyers. Indeed, those who are in negotiations to purchase land often do not even tell friends about their intentions lest their friends then secretly offer a larger amount for the property. In one case, an infirmed, elderly man with few friends in the village was suddenly besieged by visitors and free meals by those hoping to win his favor once it became known he intended to sell his prime, half-acre plot.

Not all forms of inheritance convey the ability to sell the property. What villagers refer to as "family land"—which differs from the family land described in the literature (see Besson 1987, 1988; Barrow 1992; LeFranc 1993)—is property jointly inherited by siblings after the parent-owner dies intestate. In such cases, one or more of the siblings, by joint agreement, may cultivate the land, but unless all of them agree on subdivision or sale, the land is not supposed to be sold.

What is referred to as "family land" in the literature, the villagers call "generation-to-generation land," which involves a comparatively small area of holdings, primarily house plots and land in the vicinity of houses. In this form of inheritance, the original owner indicates that rights to such land will pass to

TABLE 4.5. Distribution of Land under Various Forms of Tenure, Restin Hill

Tenure	Percentage of Land[a]
Ownership	32
Rental from private individuals	12
Crown land rental	14
Sharecropping	13
Lease	2
Family land	7
Temporary-use rights	18
Generation-to-generation land	3

Source: Survey of sixty-three of the sixty-four village households.
[a]Includes land for house plots. Percentages are rounded to the nearest whole number.

all heirs, both male and female, as well as to subsequent generations (unrestricted cognatic descent). Thus, generation-to-generation land is not supposed to be sold.[7] Ideally, those who use such land are the economically less fortunate in the family, and in the Restin Hill case, conflicts over such land were minimal.

Rental of land is particularly important in the village, but the nature of rentals varies. The most secure and least expensive type of land to rent is Crown land. The cost is only EC$10 per acre per year, and the cultivator can continue farming the property for as long as the rent is paid. Moreover, such rentals can be passed on to children. But the cultivation of Crown land does have drawbacks—much of the land is very steep, poor in quality, and distant from roads. In contrast, rentals from private landowners are much more insecure, as the agreements are on only a year-to-year basis. Yet most rentals have lasted for at least five to ten years or more, and in addition, landowners and renters are often friends. As a result, rental costs tended to be low in comparison with the expense of purchasing land. For example, in some long-standing rental agreements, annual fees range from as little as EC$60 up to EC$200 on a per acre basis. Because rentals are renewed annually, rents could increase each year—which adds a potential element of insecurity—but in many cases, rents have been fairly stable or have increased relatively slowly.

Leased land provides greater security. Generally, agreements last for five years, but prices tend to be higher per acre than those for rentals, though the cost for the five-year period is fixed. Leasing was comparatively insignificant for Restin Hill villagers, however, as landowners are generally hesitant to make such long-term commitments with their properties.

Clearly, the most contentious form of tenure is sharecropping. Two forms exist: "one-third share" and "half-share." In the former, the farmer provides all labor and inputs, while the landowner provides only the land, which entitles the owner to one-third of the proceeds from the sale of crops.[8] In the latter system, the owner not only provides the land but also pays for half the input costs, which entitles the owner to half the proceeds from the sale of crops. In both cases, the actual choice of crop is usually left to the farmer, with most preferring to cultivate bananas.

Sharecropping agreements, which are usually written, normally do not have specified periods but last for a minimum of one year. Such relationships generally tend not to last as long as rentals for two reasons. First, villagers, if at all possible, prefer not to have to give a significant portion of the fruits of their labor to someone else; indeed, the harder they work, the larger the amounts paid to the landowners. Thus, as other, more preferable forms of tenure become available—particularly by purchase or rental—people tend to terminate their involvement in sharecropping. Second, relations involved in half-share agreements can be especially contentious, for sharecroppers frequently accuse landowners of failing to contribute their stipulated portion of the cost of inputs. Similarly, landowners sometimes complain that sharecroppers cheat them out of the full amounts owed to them.

Systems based on one-third share in St. Vincent were more predominant in the past. The recent increase in the importance of half-share relations can be attributed to the rise in banana production, which involves much higher agrochemical and labor input costs than other, traditional cropping systems. Thus, farmers short of funds seek to obtain assistance from landowners to pay for half the cost of inputs, though the arrangement reduces the percentage of the proceeds they earn.

The last form of tenure involves "temporary-use rights," in which an owner grants a person from another household, usually a relative, the right to use land for an unspecified period but without formally requiring any form of compensation. However, farmers cultivating under this form of tenure usually reciprocate in some way, by occasionally providing small gifts of food, cash, or labor to the owner. In many cases, villagers cultivating such land expect, or at least hope, to own the parcels eventually through gift or inheritance.

The land tenure situation is complicated not only because of the diversity of forms of tenure but also because households are not the units that control land. Rather, land is held by individuals within households. Thus, men and women in the same household retain separate rights to their own plots; women in male-headed households have the right to determine themselves what to do on their land. The only exception is that property purchased after a couple is married is supposed to belong to both parties, though each still retains their separate interests in property obtained before marriage. Similarly, children of the household may have their own plots obtained under a variety of tenure forms.

A key issue in the literature on the Caribbean and developing countries generally is how different forms of land tenure affect agricultural production (Brierley 1987). In the case of the Caribbean, much has been written about the negative impact of what is generally referred to as "family land" (what Restin Hill villagers designate as generation-to-generation land) (see Finkel 1964; Besson 1987; Barrow 1992; LeFranc 1993). According to some of this literature, farmers supposedly hesitate to cultivate such land because all heirs, as equal owners of the property, have rights to any crops produced on it. In contrast, Restin Hill villagers assert quite firmly that such free access to any crops produced on generation-to-generation land would not be tolerated, though occasional picking of tree fruits would not be begrudged. Indeed, people cultivated bananas on such land just as they did on plots held under other forms of tenure (see also Barrow 1992; Welch 1996). Perhaps the most significant influence of tenure on land use concerns the planting of trees of economic importance, such as fruit trees, coconuts, and cinnamon. Villagers plant such trees only on land that they expect to hold for long periods, such as property that they own, family land, generation-to-generation land, and land in which they have temporary-use rights from a relative.

For banana production, however, tenure does not appear to be a significant hindrance, for growers cultivate the crop under all forms of tenure, in part because they do not have to make permanent investments on the land. Sharecroppers report that they cultivate bananas similarly on land that they own and on land that they sharecrop. Distance to roads is a more important influence on banana production than is form of tenure, given the need to carry large volumes of bananas weekly or biweekly from farms to roads.

Concern with land tenure reflects the historical and contemporary importance of agriculture in Restin Hill, the major source of income for villagers. A trip through the village reveals gardens in every possible location—from house plots to land up in the hills behind the village, the area referred to generically as "mountain." Restin Hill gardens are scattered in other nearby communities as well, just as people from other villages control land in the Restin Hill area. People usually refer to their plots by their names, often based on the name of the previous owner or on some characteristic of the area. For example, one garden was known as "self-serve" because the previous owner neglected the property, enabling others to go and help themselves to the fruits growing on the property.

Given the diverse patterns of land acquisition, land fragmentation is to be

TABLE 4.6. Distribution of Number of Plots Controlled by Agricultural Households, Restin Hill

Number of Agricultural Plots Controlled[a]	Number (%) of Agricultural Households
1	13 (25)
2	17 (33)
3	6 (12)
4	8 (15)
5	3 (6)
6	2 (4)
7	2 (4)
8	0 (0)
9	0 (0)
10	1 (2)

Source: Survey of sixty-four households, of which fifty-two were agricultural in Restin Hill.
[a]Includes only plots of 0.10 acres or larger. Percentages are rounded to the nearest whole number.

expected (Table 4.6) (see also Brierley 1987). Most households have more than one plot of land, and it is quite common for members of the same household to hold their various plots under different forms of tenure. But the degree of land fragmentation is not extreme, as the majority have only one or two fields. Fragmentation does have certain advantages. With plots at different elevations and varying soil conditions, farmers are able to take advantage of differences in rainfall related to elevation to produce a wide range of crops. Moreover, fragmentation reduces potential losses owing to pests and diseases. But banana farmers, in particular, find fragmentation a problem. Given the need to harvest weekly or biweekly, farmers with several plots are often forced to harvest from more than one garden on the same day, leading to extra time spent on travel between plots. Also, given the increasingly complex technology involved in banana harvesting and packing (discussed in Chapter 5), movement from one plot to the next creates additional burdens in relation to setting up packing operations.

The extent of land fragmentation for households varies over time, because holdings are fluid. As people obtain more land through purchase or inheritance, they tend to drop the cultivation of less desirable plots, especially those under sharecropping or Crown land rentals. Similarly, as more land becomes available closer to the village, growers will cease operations on more distant holdings. Also, those cultivating under rental or sharecropping arrangements are some-

TABLE 4.7. Distribution of Size of Agricultural Plots, Restin Hill

Size of Plot (acres)	Number (%) of Plots
0.10–0.50	74 (51)
0.51–1.00	41 (28)
1.01–2.00	16 (11)
2.01–3.00	8 (6)
3.01–4.00	1 (1)
4.01–5.00	1 (1)
>5.00	3 (2)
Total	144 (100)

Source: Survey of sixty-three of the sixty-four village households.

times forced to leave their holdings when the owner decides to sell the land or cultivate it himself or herself, adding additional fluidity to patterns.

Fragmentation of plots is not the only problem faced by farmers in general and banana growers in particular. Most plots are also small, making economies of scale in production impossible (see Table 4.7). An overwhelming majority of fields, 80 percent, consisted of land one acre in size or less, with slightly over 50 percent encompassing only one-half acre or less. The largest single plot controlled by a household was 5.75 acres.

Even a passing glance at Restin Hill and elsewhere in the Marriaqua Valley in the late 1980s and early 1990s revealed that one of the traditional concerns about Caribbean agriculture—underuse of land—was not relevant here. In Restin Hill, 78 percent of the agricultural land was under cultivation, and much of the rest was under short fallow, awaiting recultivation in the near future.[9] Similarly, another trend observed in Caribbean agriculture that raised official concern—the aging of the agricultural population and the disinterest of youth in agriculture—seemed less of a problem in the village and other banana-producing areas than was noted in the literature. Indeed, with the incomes possible from banana production in the late 1980s and early 1990s, children were willing—and in some cases quite eager—to cultivate land on their own.

Bananas were clearly the lifeblood of Restin Hill, occupying 77.3 percent of the cultivated land in 1988–89. In the village, 90 percent of the households involved in agriculture produced the crop commercially. And it was important in both male- and female-headed households. Similarly, bananas dominated irrespective of the size of holdings. Although controlling two acres of land or less influenced participation in wage labor, it did not hinder involvement in banana production (Table 4.8).

TABLE 4.8. Relation between Landholdings and Agricultural Patterns, Sample Households

Area Controlled	% of Land Cultivated	% Cultivated Land under Bananas	% Cultivated Land under Food Crops
Two acres or less	80	83	29
More than two acres	77	76	28

Source: Sample of twenty-four households.

In addition to bananas, farmers produce a variety of local food crops both for home consumption and for sale. And marketing of such crops in Kingstown is another major village preoccupation, just as it is elsewhere in the Caribbean (Katzin 1960; Durant-Gonzalez 1985; Hanseen 1990; Mintz 1989).[10] Thus, in thirty-four of the sixty-four households, at least one female marketed crops in St. Vincent regularly, with four of these households having more than one woman involved in the trade. Similarly, in 67 percent of the households that cultivated 0.5 acres or more, at least one woman marketed on a weekly or near weekly basis. Revenues from such activities varied; most Restin Hill women sold between EC$50 to EC$100 worth of produce each time they marketed in or near Kingstown.

Crop production is not the only concern of farmers, as many raise livestock as well. A walk through the village will reveal the ubiquitous chicken scurrying around, pecking in search of food; sheep or goats tethered at home, along paths, or in abandoned fields; an occasional pig; and cattle kept in either small plots called "pen patches" or on fallowed fields (Table 4.9). Villagers raise most animals for sale, the exception being chickens, which provide eggs regularly for morning meals. Similar to the situation in relation to land, rights in livestock vest in individuals, not households. Even young children sometimes own small livestock, obtained primarily by gift.

Clearly, villagers in Restin Hill were involved in a wide range of income-producing activities. Considering income from all sources—banana production, food marketing, wage labor, livestock sales, remittances, and support from companions in visiting unions—I estimate that household annual incomes of the sample ranged from EC$3,700 to EC$35,000, which indicates the widespread disparities in wealth among villagers.[11]

In research utilizing data partly from a single community, the issue of the representativeness of the study area is always of concern.[12] A recent islandwide household survey of Vincentian agriculture (Wedderburn 1995) reveals that

TABLE 4.9. Livestock Holdings, Restin Hill

Animal	Number (%) of Households Raising
Cattle	22 (34)
Sheep	14 (22)
Goats	21 (33)
Chickens	54 (84)
Donkeys	4 (6)
Pigs	9 (14)

Source: All sixty-four households in Restin Hill.

Restin Hill is very similar to the "average" community on the island in a wide range of dimensions: household size; female heads of household being older than their male counterparts; female-headed households having less land and cultivating a smaller portion of their land; limited fragmentation of holdings; ownership as the model form of tenure; the large majority of holdings being five acres or less; and the planting of tree crops on land held under more secure forms of tenure. And, as our later discussion will indicate, Restin Hill and other Vincentian communities also tend to sell the majority of their food crops (ibid.).

Another similarity concerns involvement in the banana industry. Although commercial banana production is not found in every community on the island because of distance from Kingstown or environmental constraints, it is the predominant form of land use. Thus, Restin Hill villagers share yet another situation with many others—specifically, the experience of significant changes in the nature of agricultural technology and in the extent of control by capital and the state over their labor.

Five

The Labor Question

That capital and the state have considerable influence over the peasant production process in contract farming in general and in banana production on St. Vincent in particular is clear. But much of the literature on contract farming emphasizes only the dimension of control, ignoring the manner in which farmers cope with the various constraints they face. Certainly, the concept of deskilling and the associated notion of the separation of conception and execution suggest that growers faithfully and uniformly follow the dictates of capital and the state. But the less we know about something, the more uniform it appears. Studies rarely explore, in detail, variability in production practices among contract farmers. As this examination of banana production reveals, farming requires a much more complex set of responses from growers than assembly line processes do from deskilled factory workers; farming is not a repetitive-motion enterprise. The environmental rootedness of agriculture is one of key dimensions that encourages a closer link between conception and execution than is implied in the concept of deskilling.

Moreover, the literature has ignored the crucial dimension of technological change in contract farming, which is also highly relevant to Braverman's hypothesis. I show that, contrary to the implications of the deskilling concept, technology employed in banana production has become *more* complex over time and has required increasing levels of skills, leading to a variety of impacts and responses at the local level.

Also, we cannot view control in general, absolute terms, as is characteristic in some of the literature on contract farming, because capital and the state rarely control all aspects of the production process uniformly. In relation to banana production, control has been most marked in harvesting and packing, whereas peasants retain considerable autonomy in other phases of production.

Discussions of control and the related concept of disguised wage labor necessitate consideration of the social organization of production. Banana growers need to manage a labor force, taking into consideration their household labor availability, their ability to hire workers, areas planted, and the environmental contexts of production—considerations that help to distinguish farmers from wage laborers. In addition, as technology has changed over time, farmers have had to readjust their labor recruitment patterns, an element of conception ignored in the deskilling debate in relation to contract farming.

Several political-economic forces have driven the pace of technological change at the local level. Primary among them is the need to produce high-quality fruit. Geest had clearly pressed for higher-quality bananas, complaining continually about the inadequacies of Windwards efforts. The British government added to this pressure, even threatening the Windwards with lessened market protection for their fruit if improvements were not forthcoming. At the same time, much of the financial aid provided to the industry was earmarked for quality enhancements. Adding to the chorus are the supermarkets in the United Kingdom. One of the most significant trends for the Windwards is the growing dominance of supermarket chains in the banana trade, accounting for over 60 percent of banana sales in 1994, compared with only 20 percent in the mid-1970s (Nurse and Sandiford 1995: 38–39). The growth of the supermarket trade—and its demand for prepackaging—has intensified calls for better-quality fruit, which has forced the Windwards to adopt technical innovations in harvesting and packing to continually meet the challenge (WINBAN 1986a: 15; see also Marsden 1996: 251; Watts 1996: 235). And more recently, the regulations of the European Union, representing the reality of the forces of globalization, have also mandated changes at the local level.

But the key issue of quality, which is highlighted in the literature as an important dimension of competition in the post-Fordist era, has been a crucial consideration in the banana industry even before the mid-1970s, when flexible patterns in industry were supposed to have taken hold. For example, the decline in Windwards fruit quality reported from 1961 to 1968 was the cause of much concern in the industry and sparked several technical innovations designed to alleviate the problem. A WINBAN official (Twyford 1967: 11) in 1967 proclaimed, "It cannot be too often repeated that the welfare and prosperity of the banana industry depends to a very great extent on the quality of our fruit in the United Kingdom." Similarly, an article in a WINBAN newsletter (WINBAN 1970: 2) warned:

> The recent visit to the Windward Islands of a representative of Marks & Spencer, one of the largest chain stores and dealers in fruits in the U.K. continues to remind us that the quality of our fruit is far below standard. . . . Quality standards for bananas have risen sharply in recent years and we as growers have not so far improved our techniques adequately to meet the ever higher requirements. We will have to do so as this is a world-wide trend: all quality standards for fruit and vegetables, not only bananas, are going up and in virtually every market, at home and overseas. Failure to meet the demands for better bananas, which would ultimately result in our

> marketing agents seeking high quality fruit elsewhere would be a major disaster, not only because we would have lost the most lucrative part of the banana trade but because we would have failed to live up to the challenge of the times.

Certainly, the pressures for quality improvements represent a long-term trend.

Reflecting the labor intensity of contract farming generally, banana production requires a truly Herculean effort, with a technology that is much more complex and requires more skill than any other form of agriculture ever practiced on the island. Its technological complexity and labor intensity is obvious to anyone with even a brief familiarity with operations in Vincentian banana fields. My own research indicates that in the late 1980s, banana production absorbed approximately 150 labor days per acre per year, which is much more labor intensive than traditional forms of agriculture practiced on the island.[1] And even if one does not have the good fortune to observe this unbelievably complicated process, the official WINBAN growers' manual (WINBAN 1993b) will put any doubts to rest: the most recent version is 107 pages long (though with some illustrations). The increase in complexity of the production process is reflected in a comparison with previous growers' manuals over time. One published by the SVBGA (1966) in 1966 contained only nineteen pages, another in 1981 produced by WINBAN (1981b) grew slightly, to twenty-one pages, whereas a subsequent manual increased to forty-three pages in 1986 (WINBAN 1986b).

To capture the high degree of labor intensity, the extent of variability in farmers' practices, and the technological complexity of banana production, I first discuss present-day production practices. The various steps in banana production are outlined in detail to portray the intricacies involved. What may seem like highly technical details for readers actually reflects the torturous realities of life for banana farmers. The examination then shifts to a consideration of the variable ways in which growers implement official recommendations. I next explore the issue of technological change in the industry and conclude with a discussion of the social organization of labor in banana production.

Official Banana Technology

The SVBGA, which adopted recommendations made by WINBAN, provides farmers with a standardized set of procedures. Although certain required cultivation practices are vaguely specified by statute and the Banana (Protection and Quality Control) Act of 1984 (SVG 1984) does include reference to practices detailed in the official WINBAN banana growers' manual, the Association does not strictly regulate cultivation practices—specifically, those involved in actually growing the crop—in part because it does not have adequate extension staff to monitor all farms on a regular basis. But it does attempt to control more rigidly the harvesting and packing stage by downgrading or rejecting fruit that is improperly packed or poor in quality. Thus, the extent of variability in practices among farmers is much less in the last stage of production. Before considering the issue of variability, it is worthwhile to describe official SVBGA recommendations as well as some general features of the production process.

The first step in banana production is planting. The SVBGA does not mandate the use of any one form of planting material, and the WINBAN manual lists eight different choices, some of which are discussed below. The Association recommends that farmers plant bananas eight feet apart in a square-shaped pattern that yields approximately seven hundred plants per acre; the wide space between plants is necessary because of the large leaf canopy that develops and the need to ensure that the leaves of one plant will not rub against and scratch the bunch of another.

At the time of planting, pesticides are supposed to be applied to control nematodes (*Radopholus similis*, *Rotylenchulus reniformis*, and *Helicotylenchus* spp.), tiny parasitic worms that attack the plant roots, and the banana borer (*Cosmopolites sordidus*), a black weevil that destroys the roots and corm of the plant. Subsequent nematicide applications are supposed to be made every four to six months, depending on the brand used.

Approximately three to four weeks after planting, weed control begins, with the use of herbicides generally recommended to save time and money. Farmers should weed every two months during the rainy season and at longer intervals in dry periods during the "plant" crop.[2] The weeding burden drops considerably in the ratoon stage, however, as the well-developed leaf canopy reduces the penetration of sunlight, necessitating only two weedings per year.

Shortly after weeding, fertilization begins, a task that should be repeated every two months for the life of the field. Bananas are especially demanding of nutrients; improper fertilization results in low-weight bunches, produces badly shaped fruit of inferior quality that may be downgraded or rejected, and contributes to "ship ripe," premature ripening of bananas while on the boat. It may seem that the task of fertilization is fairly easy—just sprinkling the chemical around the base of plants—but in reality, it is quite burdensome. Growers have to carry heavy 110-pound bags of fertilizer from their houses to their fields, which can be a considerable distance and over rugged terrain. And many such trips are required because farmers should apply about 2,100 pounds of fertilizer per acre per year.

Four or five months after planting, the crucial task of "follower setting" and "desuckering" (pruning) begins. Judgment here is crucial. From among the numerous suckers growing at the base of the "pseudostem," farmers must select a vigorous one that will be allowed to remain as the "follower" to produce the next ratoon crop, cutting out (desuckering) the remainder with a cutlass.[3] The official recommendation from the SVBGA is to have one follower, based on the belief that the presence of multiple followers leads to greater leaf spot problems and lower productivity.

"Pluming" fields first takes place at approximately six months after planting. The term refers to a general cleaning up of the garden and includes "detrashing," the cutting off of yellow and dry leaves hanging from the plants and removal of the peeling, dried outer covering of the pseudostems, which serves to destroy the hiding places of insects and to reduce disease levels. Done at the same time is "bunch clearing," the removal of green leaves that are likely to come in contact with and scratch the bunches. Farmers are then supposed to place the cut leaves carefully along the contours of the slope to limit soil erosion. The tasks of desuckering, detrashing, and bunch clearing should be repeated throughout the life of the field, approximately every two to three months.

Growers must now devote their attention to "deflowering." The task itself is not physically demanding: a farmer simply uses his hand to brush off the flowers growing at the end of each "finger" (an individual banana). But timing is crucial, as the task should be performed when the developing fingers of the bunches are horizontal to limit problems with dripping latex. Unfortunately for farmers, they cannot deflower an entire field at once because of different rates of maturation among plants in the same field; rather, the task often extends over several months and, in the ratoon stage (when the plants are taller), requires use of a ladder, thus consuming considerable time.

A few days after deflowering, most farmers follow the recommended practice of "sleeving," covering their bunches with a thin, blue diothene (polyethylene) bag called "tubing."[4] Sleeving reduces insect damage, increases bunch weight, and shortens maturation time, though it also makes bananas more susceptible to bruising. Similar to deflowering, the task becomes more burdensome in the ratoon stage, when the taller bananas require use of a ladder. Sleeving itself is relatively easy, taking just a little more than a minute or so to cover each bunch, but it can also be quite hazardous when using a ladder on steep hills, where growers often have to drape one leg around the pseudostem while deflowering or sleeving to prevent a fall downhill.

The next step, "propping," which is intended to prevent the top-heavy plants from falling over, is crucial, for bananas are easily blown down by the strong winds that frequent the region. In addition, soils saturated by continual rainfall make these shallow-rooted plants more likely to topple, especially on steep slopes. The recommended practice, which provides the most secure anchoring, is to tie one end of a length of twine to the top of the banana plant and fasten the other end to a short wooden stake driven into the ground. Although providing considerable stability, the method requires climbing on a ladder for each bunch as well as cutting wooden stakes, which adds to the labor burden. Propping is not a major endeavor in the shorter plant crop, which usually ranges between 85 inches and 95 inches in height, unless the slope is very steep, which encourages the plants to lean downhill. In each successive ratoon, however, the plants grow taller, normally ranging between 110 inches and 126 inches (depending on the cultivar), as they continually compete with one another for sunlight.

The final task is harvesting. Given the large amounts that can be produced from peasant plots, it is also the most labor-demanding aspect of banana production; my own data, based on fifteen households' sales to the SVBGA over a twelve-month period, revealed marketable yields ranging from 3.4 to 11.0 tons per acre per year, with an average of 7.0 tons. Indeed, many farmers prepare for the arduous effort by resting or working lightly the previous day in anticipation of the rapid, intensive pace of activity the next. Little latitude is possible in the timing of harvesting. Bananas ready for market one week may be overgrade and thus unsuitable the next. Given continual production throughout the year, the SVBGA has to buy bananas on a regular basis; it devotes two days each week, now Mondays and Tuesdays, to the endeavor.[5]

Yet the actual labor burden in harvesting at any point in time can vary considerably for farmers, depending on the stage of production of their various fields and the season of the year. The original plant crop usually produces the smallest yield.[6] Output in the first ratoon is highest, yielding approximately 20 percent more than the plant crop, while harvests in the second ratoon normally decline slightly from that of the first. Output usually tapers off in the third ratoon as farmers tend to provide less care to ratoons than to the plant crop. Rainfall conditions also affect harvests. In the dry season, plants mature and fruits fill more slowly compared with in the rainy season.

The current method of harvesting and packing bananas, called "cluster packing" or "mini–wet pack," is relatively new to the Windwards, having been uniformly adopted in 1993. The process begins early in the morning at home. Having received from the SVBGA the previous day their empty cardboard boxes used for packing bananas, villagers gather together the number that they expect to need, usually tying them together to make transport easier. With some carrying—or, put more accurately, balancing—as many as twenty-seven boxes on their heads, they travel up winding, narrow trails to their fields; given the large volumes that can be produced, banana growers sometimes have to make repeated trips back and forth from homes to gardens just to bring a sufficient number of boxes to their fields. They must also haul a ten-gallon plastic tub, harvesting trays, a hanging scale, a small packet containing a fungicide, and their cutlass and knife, along with five to ten gallons of clean water, if a source is not available close to their garden.

Harvesting itself takes considerable skill, and any step performed incorrectly can lead to downgrading or rejection of fruit. To begin the procedure, a villager first makes a short, vertical cut or notch in the upper portion of the pseudostem using a razor-sharp, hook-shaped knife called a "grading" or "dehanding" knife, which enables him to lower the bunch to within reach (each pseudostem produces one bunch, and a bunch in St. Vincent usually contains between six to ten "hands" of bananas).[7] Then, he lifts up the thin, blue diothene "tubing" to expose the bunch, being careful not to rip it, as it can be washed and reused several times. With a swift, short stroke of the grading knife across the stalk of the bunch, the grower cuts off or "dehands" the first hand of bananas at the bottom of the bunch. Precision is then required in cutting away the remaining small piece of stalk still attached to the "crown" of the severed hand, as a smooth surface must be created. While cutting bananas and carving the crown, the hand of bananas must be positioned properly to prevent the milky and sticky latex, which readily drips from the cut stalk and crown, from blemishing and staining the peel of the fruit. Now the villager subdivides each hand into clusters, which must contain between four and nine banana fingers. Selecting exactly where to cut a hand to produce the clusters requires careful inspection, as blemished or malformed fingers must be discarded. With latex continuing to ooze from the crowns of the clusters, the grower places each one on the midrib of a clean, freshly cut, green banana leaf set on the ground, allowing the liquid to drain for ten minutes. While the villager continues the steady pace of harvesting from the bunch, others load the drained clusters onto specially designed plastic trays or plastic boxes, carefully placing strips of soft, green banana leaves or blue diothene between each cluster as cushioning to prevent rubbing and bruising of the fruit; the lightweight plastic trays, measuring twenty-eight inches by twenty-eight inches, hold roughly fifty pounds of fruit, while the larger and deeper rectangular plastic boxes can hold sixty pounds or more of fruit. When the tray or box is filled, they carry the fruit on their heads to the central processing and packing area. The task of transportation is made arduous enough by the heavy weight of the fruit, with carriers requiring assistance to lift loads onto their heads. But even more difficult is navigating through the often steep, slippery banana fields, a task requiring vigilance and luck not to run into the often difficult-to-see twine used for propping bananas, fall into an occasional unfilled banana hole, trip over a fallen banana plant, or slip on a discarded, rotting leaf or banana. After delivering the bananas, the carriers must then return to the garden to pick up more harvested fruit and cut new green banana leaves to pack among the clusters, as the other material used on the preceding trips must be discarded.

Processing and packing of bananas take place in a central location in or near the fields. The official recommendation is for growers to build a small, open-air shed where they can wash and pack the bananas and keep the boxes out of the rain. Such structures, which can be seen dotting the landscape, are made mostly of wood and bamboo walls and a galvanized roof and measure usually from fifty to one hundred square feet in floor area.

At this location, other workers unload the clusters from the trays and plastic boxes, placing each one in the ten-gallon tub filled with either five or ten gallons of water (depending on the amount of fruit to be processed), to which they add the powdered fungicide Imazalil. They meticulously wash off any dirt, latex spots on the fruit, and insects—such as wayward ants, cockroaches, and spiders hiding among the fingers of the clusters. After allowing the clusters to remain in the tub for a few minutes, during which time the Imazalil treats the crown to help prevent crown rot, they place them on the table in the shed, where the fruit are allowed to drain briefly.

The packer, who is usually female, is now ready to begin. Packing requires considerable care, as improperly packed bananas can lead to bruising and rejection of fruit. The packer first puts a brown craft paper liner into the box, on top of which is placed a clear, polyethylene bag to prevent the bananas from rubbing against one another and the sides of the box and to help retain moisture. Now the fruit is ready to be boxed, but not all bananas can be placed in the same cartons, as short-fingered bananas, those between 5.5 and 7.4 inches in length, must be boxed separately from larger fruit, according to recently issued EU regulations. Selecting the clusters as they are placed inside the boxes requires patience and a keen eye. Packers intently inspect the bananas available on the table to make sure that they choose those which are of appropriate, uniform size for the row being packed and which will fit snugly to prevent subsequent movement. Indeed, it is not uncommon for the women to place clusters into the box only to remove them again because they were not just the right size or shape. After finishing placing the first row of clusters into the box, the packer then adds another two or three rows of fruit on top, folding the craft paper liner and polyethylene bag over each successive row to serve as protection. For regular-length fruit, a villager usually packs between fourteen and seventeen clusters in a box. When finished, she ties the top of the polyethylene bag, places the cover on the box, and affixes a sticker containing the grower's identification number.

But all is not done. Now each box must be placed on a hanging scale, which has a red mark on it to indicate the exact weight required—thirty-seven pounds (thirty-four pounds net weight), again reflecting new EU regulations. Boxes weighing more or less than that specified will be rejected. Such precision makes the task of packing even more impressive, as villagers have to make sure not only that the clusters of bananas are the right size and shape for each row but also that the total weight will be thirty-seven pounds. An incorrect weight requires repacking of the box. Indeed, many packers have developed such a fine sensitivity to the proper weight of a box that they can tell if one is the correct weight by just picking it up and holding it. With the task of packing finished, the boxes are ready to be transported.

The SVBGA pays private truckers to pick up the banana boxes, and thus people have to carry or "drogh" the boxes from the packing area to a motorable road.[8] Men usually carry between two and three boxes, though a few carry as many as four, quite a feat given that each one weighs thirty-seven pounds. Women and children normally carry between one and two. Fortunately for most villagers, given the relative location of fields and roads, most people can carry their boxes downhill. Nonetheless, the task is still difficult, for people must balance the boxes on their heads while negotiating narrow, winding, and often slippery dirt paths. After an exhausting day, household members, accompanied occasionally by friends and wageworkers, return home, with men especially looking forward to visiting rum shops to "cool out."

The exact number of people working together in harvesting and packing, along with their levels of productivity, can vary considerably. One family of four whom I observed in 1995 worked from 8:40 A.M. to 12:50 P.M. and packed and carried to the road thirty-one boxes from their half-acre banana field, with the father dehanding the fruit, his adult daughter doing the dipping/washing and packing, and two sons carrying bananas.[9] Another, larger group, composed of husband and wife and eight employees, worked varying times between 8:30 A.M. and 12:55 P.M. (for a total of thirty-seven hours, ten minutes), producing fifty-six boxes from a steep banana parcel of one and a half acres, with two women packing, one dipping/washing, four male droghers, and three men dehanding.

But not all growers perform each task as recommended by the SVBGA. Farmers modify the outlined procedures to cope with their labor constraints, the need to integrate bananas and food crops in the same fields, the small size of holdings, and their highly variable set of environmental resources—reflecting the influences of slope, soil type, elevation, likely rainfall, and site exposure to winds. And for some tasks, such as selection of planting material, the SVBGA leaves the choice up to the growers. It is instructive now to examine the highly variable nature of grower practices, as it bears directly on the issues of control over labor and deskilling.

Variability in Banana Production

Villagers indicate that while the recommended square-shaped planting pattern is certainly suitable for relatively flat land, it is not appropriate for their mostly steep plots. If they followed the recommended procedure in such an environmental context, a falling banana plant could easily set off a disastrous chain reaction by knocking down others aligned with it downhill. Thus, growers devise their own seemingly haphazard planting patterns to cope with the problem of steep terrain.

At the same time, farmers have to select the appropriate planting material, a decision that reflects a multitude of considerations. They must choose a "bullhead," the bulbous corm at the base of a plant that has already produced a bunch; a "sword sucker," which is a sucker or offshoot that sprouts from the side of a corm and is about one to two feet tall with small, narrow leaves; or a "maiden," which is a taller sucker about five to six feet in height. Each has advantages and disadvantages. Bullheads are the least favorite choice in Restin Hill. Although they produce the highest yields of all planting materials, they also take longest to mature, with a bunch becoming harvestable in ten or more months; bullheads are susceptible to waterlogging as well. In contrast, farmers prefer to plant sword suckers, which yield a harvestable bunch in only nine months and which are not as susceptible to waterlogging. Similar to bullheads, they can also be interplanted with food crops, a major advantage for many growers with limited holdings. Maidens, the tallest planting material, cannot be interplanted with food crops because they would provide too much shade before the food crops are harvested. But if farmers need cash in the short term, maidens are ideal—they yield a harvestable bunch in only six or seven months, though they tend to produce the lowest yields of all planting materials. Furthermore, maidens have the additional advantage of making weeding easier; these taller plants provide shade more quickly than do the others, which reduces the subsequent labor burden of weed control.

Pesticide use at planting also varies from official recommendations. Many growers interplanting bananas with food crops delay application of nematicides until nine months after planting, by which time their provisions are harvested, fearing that earlier applications as suggested by the SVBGA would result in contamination of their food crops. Others will apply such pesticides at planting but then wait until nine months later before reapplication, for the same reason.

Weed control also requires judgment. Farmers rely primarily on the herbicide Gramoxone for the task, but exceptions are evident. At the time of first weeding, some will hand-weed if the bananas are intercropped with food crops, to prevent the spray from damaging their other plants. Also, if the dry season is prolonged and intense, they will also hand-weed near the base of plants but leave the remaining weeds elsewhere in the field undisturbed in the belief that the presence of such weeds keeps the ground cool and preserves moisture.

Variability is also evident in something as seemingly straightforward as fertilizer applications. Given that growers obtain most fertilizers on credit from the SVBGA, the Association expects them to apply these inputs only on bananas. But the method of first application varies according to the crop mix. In fields interplanted with food crops, villagers apply fertilizers the first time only to their food crops; otherwise, the bananas would grow too quickly and provide too much shade, inhibiting growth of the food crops. Even methods of measuring dosages vary; some dispense fertilizers using their bare hands, estimating the amounts applied, while others are more careful, using more precise measuring devices, such as cups.

One of the major concerns that the SVBGA has about cultivation practices is the peasants' failure to follow the suggested pattern of retaining just one follower per plant. In contrast, villagers prefer to allow two or more followers to survive to maximize output from their small parcels. Concerned about labor constraints as well, they note that the denser leaf cover and resulting more complete shade provided by the leaves of several followers reduce the labor burden in weed control.

Even disposal of banana leaves cut during pluming and bunch clearing is not uniform. Only a few follow the SVBGA suggestion of aligning the leaves across the slope to reduce erosion. But growers do not simply toss them away. Rather, their placement reflects perceptions of the needs of their fields. Some put them near the base of their plants on the downslope side to limit the extent to which runoff can wash away costly agrochemicals. Another strategy is to toss the broad leaves on top of growing patches of weeds to suppress them. Such debris is also an effective mulch, with some villagers carefully scattering them as uniformly as possible throughout their fields in times of limited rainfall to preserve soil moisture.

Propping, too, exhibits considerable variations. The recommended practice requires climbing a ladder to tie one end of a length of twine to the top of each plant, which is time consuming. One ingenious alternative that villagers have devised involves tying the twine around a short stick, throwing it over the top of the plant, pulling it back until firmly stuck in the unfurling leaves, and then fastening the other end to a stake in the ground. Although less secure than tying twine directly to the top of the plant, it obviates the need to use a ladder. Another, even less sturdy practice is to forgo the use of stakes driven into the ground altogether and simply to tie one banana plant to the pseudostem of another. In deciding which method to employ, farmers must balance the need to save time with the problems presented by the steepness of their garden slopes and the heights of their plants, which grow taller with each successive crop; the taller the plants and the steeper the slopes, the more secure the propping that is required.

Similarly, many growers have not built packing sheds as required by government statute. Those producing bananas on land that they rent or sharecrop are hesitant to make such permanent investments on property that they do not own. And farmers with many small, scattered plots also find it unfeasible to build numerous sheds. Villagers without such structures simply carry a tarpaulin or large plastic sheet to their gardens to protect their boxes from the rain.

The scheduling of harvesting is left up to the farmers themselves. Although the SVBGA purchases bananas two days weekly, most farmers harvest their crops only one day every other week to limit their labor expenditures. But circumstances sometimes force changes in scheduling. Those with more extensive holdings have to harvest each week—and sometimes even two days per week—to cope with the large volumes produced. Even those with small areas may have to harvest each week if they have limited amounts of labor available. Similarly, farmers may be forced to harvest weekly if their banana plots are widely scattered.

The lack of uniformity and the disregard of official production practices during cultivation is certainly not new. A report (Henderson et al. 1975: 5) based on a survey of growers in 1975 revealed that only four of the twenty-two practices recommended by WINBAN were adopted by 90 percent or more of the banana farmers in all the Windward Islands. While the notion of deskilling implies separation of conception and execution and uniform adoption of work routines, such a characterization is certainly inapplicable to banana production. Growers creatively use their own initiative to cope with their particular labor constraints, small size of holdings, highly variable environmental contexts, and the integration of local food crops and bananas into the same production system.

The Evolution of Technology in Banana Production

The literature on contract farming has ignored the role of changing technology. But if we want to examine whether the labor process has become increasingly simplified, as Braverman suggests, we require a more thorough understanding of technological change and the labor process. In the case of the banana industry, technology has become more complex; moreover, the pace of technological change, especially in harvesting and packing, has accelerated over time, requiring continual skill upgrading on the part of farmers, an outcome that questions the applicability of the concept of deskilling.

Although documenting the nature of technological change, in itself, is important, we must also understand why it has occurred. Technological change does not take place in a vacuum. Technology is introduced to solve problems, and how those problems become defined and how their solutions are formulated in relation to resource-use patterns are crucial considerations in political ecology (Blaikie 1995a). In the case of the Windward Islands banana industry, officials of the British government, WINBAN, and Geest all defined the problems as low profitability in banana farming and limited international competitiveness. The factors believed responsible for these problems were poor fruit quality and low levels of farm productivity. The solution was defined as being technological in nature: if only the Windwards growers would adopt the appropriate technologies and practice them faithfully, productivity and quality would increase, which would then improve profitability and competitiveness. Because the solution was defined primarily in technological terms for much of the history of the industry, officials paid less attention to the constraints on profitability that we have already examined, especially the unfavorable contracts with Geest. In fact, I would argue that if the Windwards growers had received a larger share of revenues from the sale of their bananas, technological innovations would have been much more successful in achieving the intended goals. My own observations in the late 1980s and 1990s indicate that when their returns were higher, as they were in the late 1980s, farmers' cultivation practices were more conducive to improving both fruit quality and productivity—they applied agrochemicals, attended to weeding, and performed other suggested field practices, such as pluming and sleeving, on a much more regular basis than they have been in the mid-1990s, when profitability declined dramatically.

The point is not to suggest that the Windwards have had no problems with either fruit quality or farm productivity. And certainly some technological changes have clearly benefited the growers. Rather, I assert that attempting to solve the impacts of an inherently inequitable relationship with Geest by relying primarily on innovations in cultivation, harvesting, and packing ensured that the Windwards growers would be condemned to experience a seemingly endless cycle of technological change. And such technological change alone will never be a sufficient solution to the problems of the Windwards banana industry.

In the following discussion I first consider changes in cultivation practices and then discuss those in harvesting and packing. It is in the realm of the latter that capital and the state have been able to exercise the most control. It is also the realm that has experienced the most rapid rate of change.

Cultivation practices were much less complex and less labor intensive in the early days of the industry in the 1950s compared with today. Growers did not deflower or sleeve their bananas. They did apply fertilizer, which at that time came in bulky 224-pound bags, but in lesser amounts. Villagers rarely used pesticides themselves, as pest problems were fairly minimal. Weed control was by hand, with farmers employing a method called "circle weeding," in which they used a hoe to clear weeds carefully at the base of the banana plants in a circular area, while just trimming back weeds with a cutlass elsewhere in their fields. Although circle weeding was very labor intensive, farmers did not attempt to control weeds as frequently then as they do today. Similarly, their effort at pluming was minimal. For propping, farmers used a bamboo pole wedged against the top of the pseudostem.

Change was incremental in the 1960s. Farmers slowly began to use more pesticides, such as aldrin and later heptachlor, to help control the increasing depredations of the banana borer and, much more rarely, Nemagon (DBCP, or dibromochloropropane), a foul-smelling pesticide injected into the soil, to kill nematodes. Other practices remained the same as in the 1950s, with a slight increase in the use of fertilizers.

During the first half of the 1970s, farmers devoted even less attention to their fields compared with the 1960s. Prolonged drought, plummeting banana prices, and skyrocketing fertilizer costs made banana production marginal at best for many and drove others out of the industry altogether. Villagers returned to banana production and increased devotion to their fields in the second half of the decade, spurred by improved banana prices, a growing agricultural extension effort, and several British-funded aid programs, especially the Banana Development Program, which lasted from 1977 to 1982 (see Grossman 1994). In the mid-1970s, use of the herbicide Gramoxone slowly began to replace hand weeding, and by the late 1970s, growers had switched to twine to prop bananas and had also begun sleeving their bunches. British-funded subsidies for agrochemicals, both fertilizers and pesticides, contributed to more frequent applications. Growers also began to devote greater attention to pluming.

Although farmers were adopting new practices and devoting more time to traditional ones, such as weeding, in the 1970s, some changes improved labor productivity. According to villagers, weed control with Gramoxone takes only one-third the time compared with hand weeding; moreover, weed regrowth is slower when herbicides are used. Similarly, propping with twine requires approximately one-half to one-third the time as did the use of bamboo. Not only did growers no longer have to cut, clean, and transport the bamboo poles to their fields, but twine also lasts longer—it can be reused for a few years, whereas bamboo poles rot in as short a time as six months. In addition, if used properly, twine provides more secure support.

Reflecting even higher banana prices and a more determined extension effort, growers intensified field care in the 1980s, paying greater attention to weed control and field sanitation. Many began using nematicides for the first time, as pest infestations escalated. Growers also had to start deflowering with the advent of "field packing," a revolutionary form of harvesting and packing that I discuss below. Although sleeving had been introduced earlier, its use became much more widespread in the 1980s.

By the end of the 1980s, cultivation practices had become more complex and required more skill compared with previous decades. Technological change did result in improvements in its intended targets—fruit quality and farm-level productivity. Although it is not possible to compare quantitatively improvements in fruit quality over time because of the different grading systems used, it is clear that quality improved considerably over that in previous decades; what was once acceptable fruit quality in the past is no longer suitable for the market today. Similarly, production per acre had increased.

The pace of change in harvesting, packing, and transporting green fruit by villagers has been even more rapid. Unless great care is taken at this stage of the production process, quality will suffer, irrespective of the attention devoted to cultivation practices, because improper packing of bananas leads to bruising and scratching of the fruit. Furthermore, it has been necessary to devise suitable methods of packing and transporting to cope with the potentially rough treatment that fruit can experience as farmers carry their bananas over steep, rugged terrain and trucks transport them over poor, winding roads dotted with numerous potholes.

What is particularly interesting about this phase of production is the gradual shifting of processes once performed off the farm onto the growers themselves. Thus, it is necessary to describe certain relevant off-farm procedures in both St. Vincent and the United Kingdom.

In the 1950s, harvesting was a fairly simple affair. Growers just cut whole bunches and carried them horizontally—usually two or three at a time—on their heads. The only items used to preserve quality were banana leaves. A small, circular pad made of rolled-up soft leaves served as cushioning for growers' heads and helped them balance their loads. To prevent the bunches, which were tied together, from rubbing against each other, villagers cut off the dead, dried leaves hanging from the pseudostems in their fields, moistened them with water to make them soft, and placed them between the bunches to serve as padding. Then, they carried the green fruit to SVBGA buying stations—small, open-air structures with galvanized roofs—where SVBGA personnel would inspect and weigh the bunches. The Association, in turn, transported the whole bunches in trucks, along with banana leaves layered among them to serve as padding, to the port in Kingstown for export. Workers at Geest in the United Kingdom ripened the fruit and subsequently had to cut individual hands off the bunches, trim the crowns of the hands, and remove the dried flower attached to the end of each finger before selling them to wholesalers and retailers.

Carrying banana bunches in a horizontal position with minimal protection made the fruit highly susceptible to bruising, and Geest complained increasingly about the poor quality of bananas being exported. To reduce the problem of bruising, in the 1960s the SVBGA introduced a flat, padded wooden tray, approximately twenty-four inches wide by thirty-six inches long, which growers were supposed to carry on their heads to transport bananas. In addition, farmers were encouraged to wrap each individual bunch in a sheet of foam for additional protection. Farmers adopting these practices received a small bonus from the SVBGA, though they had to purchase the half-inch foam sheets from the Association.

The extent to which villagers adopted the new practices was highly variable, but only a small percentage used both the padded tray and foam, because farmers felt that the added costs were simply not worth the minimal benefits. Most continued to carry bananas as they had previously. Others used the flat trays alone or purchased only the foam to wrap the bunches. In any case, the innovations were of limited benefit: continued transportation of bananas in a horizontal position encouraged bruising, even with the addition of foam padding. The basic problem was that the flexible banana stalk acted somewhat like a spring when carried in a horizontal position, with the resulting up and down movement subjecting the bunches to bruising (Sealy and Hart 1984).

The SVBGA also tried to reduce bruising of fruit occurring in its own operations. In the latter part of the 1950s, it began wrapping each bunch in cushioned paper and diothene, the latter helping fruit retain moisture. In addition, the Association required trucks carrying bananas from buying stations to Kingstown to pad their side walls with foam and place foam pads between every layer of horizontally resting bunches.

Such incremental change represented by the use of padded trays and foam and wrapping bunches in cushioned paper and diothene did not result in significant improvements. Desperate to improve quality, the Windwards began erecting "boxing plants" to process and pack bananas in cardboard boxes before export. The SVBGA built its first plant in 1968, and by 1971 it was processing all fruit at these stations. Located throughout the island, boxing plants were large, warehouselike structures. The one in LaCroix, which serviced the village of Restin Hill and the surrounding area, employed approximately fifty people, most of whom were women. Once growers delivered their bunches, workers at the boxing plants weighed, inspected, graded, and dehanded the fruit. After trimming the crowns of the hands, workers processed the green bananas in two large, rectangular, water-filled tanks, where they deflowered, washed, and treated the bananas to stop the flow of latex. When they were finished, the fruit received a fungicide treatment to prevent crown rot. Finally, employees packed the individual hands into cardboard boxes, which the SVBGA then shipped by truck to the port in Kingstown for export. Called the "wet-pack" method, this new processing system relieved Geest in the United Kingdom of the tasks of dehanding bananas, trimming the crowns, deflowering them, and packing the bananas into boxes before sale—thus reducing the company's labor expenses.

The Windwards did not obtain the boost in fruit quality expected from the transition to boxing plants. To improve the situation, the SVBGA focused once more on changing harvesting and packing procedures. Starting in 1972, it encouraged farmers to terminate the old methods of carrying whole bunches horizontally. Rather, it urged them to adopt the use of hard, plastic "field boxes," which measured 28.0 inches long, 11.5 inches wide, and 10.5 inches tall. In this new system, growers were supposed to carry harvested bunches by the stalk to a centrally located place in their own fields and dehand them there. Then, they rested the individual hands on a freshly cut, green banana leaf to allow the latex to drain from the crowns for a few minutes. From the harvested banana plants, they also had to cut off numerous green leaves; discard the stiff, central midribs; and place the remaining shredded, soft leaves in between the green hands as cushioning while packing them into the plastic boxes. Villagers then had to carry the bulky boxes on their heads to a motorable road, where truckers hired by the SVBGA picked up the boxes and transported them to the boxing plants, at which point SVBGA workers processed them using the wet-pack method. Thus, a key element of processing once performed off-farm—dehanding—was now accomplished in the field.

By the mid-1970s, most growers in St. Vincent were using the plastic field boxes, but people in Restin Hill and elsewhere in St. Vincent and the Windwards altered the method. The SVBGA recommended that farmers fill the plastic boxes with only one row of banana hands to minimize the bruising and scratching of fruit, advice that was almost universally ignored. Packing one row of fruit enabled growers to carry only thirty pounds at a time, whereas filling them with two rows yielded heavier loads of approximately seventy pounds (Felton 1981b). Consequently, villagers generally packed two rows in each one to minimize by one-half the time spent transporting their fruit.

Although more labor intensive than previous methods, the new system did have a very important advantage for farmers. Rejection rates dropped. In one of the boxing plants that I visited, an old poster on the wall extolled the virtues of the system of plastic field boxes, indicating that rejection rates for fruit transported the previous way—as whole bunches carried on the heads of villagers—

were as high as 25 to 40 percent, whereas with the use of plastic field boxes, such rates dropped to 10 to 15 percent.

Although the quality of fruit delivered from the fields improved, the boxing plant system itself created problems. At the plants, the extensive handling of fruit, coupled with low-waged workers having similarly low levels of motivation, resulted in considerable bruising and scratching of bananas. At the same time, pressure from Geest and the British government continued to mount for improvements in Windwards fruit quality.

The Windwards thus embarked on yet another transformation of harvesting and packing practices. The revolutionary new system, called "field packing," was developed jointly by Geest and WINBAN and was practiced nowhere else in the world, except the Windwards. Although a few farmers started field packing in 1981, it was not until 1986 that all fruit was processed in St. Vincent using this method. The system greatly increased the level of complexity and skill required in banana production. Moreover, it moved even more tasks previously performed off-farm to the farmers themselves—deflowering, trimming of crowns, crown rot control, and boxing. While increasing the labor intensity of banana production, it also increased farmers' incomes—they received an extra three cents per pound for packing the bananas and an additional one cent per pound bonus for deflowering. Furthermore, the resulting improvements in fruit quality led to higher prices received from Geest.

In field packing, unlike in the previous system using plastic field boxes, all work was done directly in gardens next to the harvested bananas. In addition, much more care was required. The system of harvesting and trimming of the crowns of each hand was the same as that employed today in the cluster pack method, except that farmers did not cut each hand into clusters and allowed the latex to drain from each hand for only two to five minutes instead of the ten minutes necessary today. In particular, they had to make sure that the oozing latex did not come in contact with the fingers of the hands because they could no longer rely on the boxing plants to wash off the substance, which leaves a cosmetic defect. The next required step was to place a small (4.50 by 2.75 inches), green, rectangular-shaped, fungicide-treated (thiabendazole) pad on the crown of each hand to prevent "crown rot," a perennial problem in the Windwards. Here, two previous steps were critical. Unless the crown was trimmed smoothly, the pad would not stick properly. Also, if a grower allowed latex to drain from the hands for more than the recommended time, the nature of the latex would change such that the fungicide in the pad would not be activated properly.

Now growers were ready to pack the bananas. The cardboard boxes, measuring twenty-one inches long by fourteen inches wide and almost eight inches tall, were subdivided into two internal compartments to hold the hands. Villagers first had to position a piece of thin, transparent diothene evenly into the two sections to prevent the hands from rubbing against the sides of the box. Then, they carefully placed three or four hands into each of the two compartments, making sure that the crown of one hand did not scratch or bruise the fruit below it. After filling the box, they folded the diothene over the top of the bananas and placed the cover on top. Finally, they carried the packed cardboard boxes on their heads from field to road, where truckers took them to the old boxing plants, which had now been converted into inspection stations called IBDs.

Field packing was particularly significant because it represented a qualitative transformation in the relation of the peasantry to capital and the state. Previously, capital and the state did not mandate that peasants had to use a specific form of harvesting and packing, though the SVBGA encouraged adoption of innovations through extension supervision and the payment of small bonuses. But now, unless peasants delivered bananas already packed in the cardboard boxes according to SVBGA specifications, the Association would reject the fruit as not suitable for export. Similarly, subsequent innovations in this phase of production have all eventually become mandatory as well. But what is especially interesting is that such increased control did not lead to deskilling. On the contrary, the skills required by field packing and subsequent mandatory systems have been more substantive and complex than those associated with previous methods of packing.

Field packing clearly increased the labor intensity of production, but farmers uniformly praised it. Most specifically, it led to improvements in fruit quality, which helped lower the rejection rate to approximately 4 or 5 percent in the late 1980s; the reduced handling of fruit compared with previous systems was particularly important in enhancing quality. Moreover, farmers found that the compact cardboard boxes were also easier to carry and that they could carry more bananas at one time than with the old system using plastic field boxes. In addition, the amount of time that they had to spend waiting at the IBDs to get their bananas inspected was less than the time that they had had to spend while their fruit was being processed through the previous "wet-pack system" at the old boxing plants.

Even with significant increases in fruit quality, neither the British government nor Geest was satisfied with the results. Indeed, in the 1980s, the British government began to set specific targets for improvements in fruit quality, once again threatening to allow increased importation of Latin American fruit if the Windwards failed to meet the objectives it established. And in the late 1980s, Geest argued that a new system of harvesting and packing, "cluster packing" (which is now used in the Windwards), was necessary to improve quality and marketability. The marketing issue became particularly important to Geest because of possible changes that could occur with the proposed Single European Market, which would create a much more competitive environment for the Windwards. Specifically, Geest argued that the Windwards were the only producers in the world marketing bananas in the form of whole hands and using crown pads, while everyone else sold bananas in the form of clusters.

Arguing that cluster packing was too labor intensive, WINBAN initially resisted using this method. Instead, it devised a new method that it called "paco pack," which it introduced to most Windward farmers, including Vincentians, by 1990. Now growers had to place a long, craft paper liner into each of the two compartments in the cardboard box and position a polyethylene bag on top, both of which they folded over each successive hand packed into the compartments. The system was soon modified in 1991 at the insistence of Geest, which argued that other producers selling bananas in Europe did not have internal compartments in their boxes. Thus, to make their boxes appear more similar to those of others and hopefully improve marketability, the Windwards removed the central divider in the box, which also enabled villagers to pack slightly more in each carton.

Still, Geest was not satisfied, pressing intently for the adoption of cluster packing. It complained that many boxes it received were "underweight"—that is, they were not fully packed—a problem it attributed to the packing of full hands. Keeping its focus on the upcoming Single European Market, it also argued that Windwards bananas would be at a competitive disadvantage because of the use of crown pads. Although British buyers of Windwards bananas were used to the unsightly dark stains that developed on the pads, potentially new customers elsewhere in Europe might find them unappealing. Geest also claimed that the crown pads were becoming increasingly ineffective in controlling crown rot, thereby reducing the quality of Windwards fruit.

Finally succumbing to the pressure, WINBAN introduced the current harvesting and packing system—"cluster packing" or "mini–wet pack," described previously in this chapter, and by 1993, all bananas exported from St. Vincent were being processed using this method. The system, which was a Jamaican innovation that WINBAN borrowed, was supposed to have several benefits. Besides removing the unsightly crown pad and marketing the fruit in the form of clusters, it was also supposed to limit bruising during the packing and transporting of fruit and to reduce crown rot infections. However, quality has fluctuated markedly since cluster packing was introduced (Figure 2.2).

But the system has also radically increased labor burdens for farmers. Cluster packing demands additional time in relation to cutting hands into clusters and transporting the bananas from where they are harvested to a central processing location (instead of packing them next to the harvested plants, as was done previously). Washing bananas and liquid application of a fungicide to control crown rot are much more time consuming than the use of crown pads, but they have become a necessity, because crown pads can be applied only to full hands, not clusters. Indeed, some wash the bananas in such a painstaking manner that they almost appear as if they are preparing the clusters for surgery. Women are also forced to take more time and be more selective in packing; in packing clusters, which are smaller than full hands, they must ensure that each one is the appropriate size and shape for the row being filled. In addition, packers often experience "dead time"—idly waiting for more bananas to arrive after having finished washing and packing the fruit already delivered. And, once again, another task that was previously performed off-farm has been pushed onto the growers: cutting hands into clusters was the task of Geest employees in the United Kingdom, whose services were now no longer needed.

But more demands besides those of Geest were soon to follow. In 1995, new EU regulations governing the marketing of bananas in the Single European Market forced villagers to begin packing short-fingered bananas in separate boxes. Furthermore, all boxes now had to be weighed by the villagers themselves, another task previously performed off the farm by the SVBGA.

Certainly, labor intensification in the banana industry is not new. Using plastic field boxes, field packing, and paco pack demanded successively increased labor inputs, but growers found them useful and worthwhile innovations. In contrast, the Restin Hill reaction to cluster packing, similar to that throughout the island, has been dramatically different—almost uniformly negative. Indeed, a recent WINBAN (Alexander-Louis 1992) study based on a small sample in St. Lucia reported that cluster packing requires 31 percent more labor in harvesting and packing than did the paco pack system. But because of changes in the social organization of labor, the increased burdens for growers have been even greater (see the next section).

These changes may seem incremental to the outside observer, but they were certainly not to the villagers. For example, the average amounts packed in each system gradually increased in net weight from 28 pounds in the field pack system to between 29.5 and 32 pounds in the paco pack systems and to 34 pounds in cluster packing.[10] Also, cardboard box sizes in each system were slightly different, requiring the adjustment of fruit selection strategies in packing. Moreover, the permissible range in the amount of fruit packed in each box has dropped considerably since the introduction of field packing, requiring more precise positioning of fruit in the boxes. Whereas under the field pack

TABLE 5.1. Duration of Systems of Harvesting and Packing Bananas

System of Harvesting/Packing	Dates Most Widely Practiced[a]	Years
Full bunches on the head	1954–1975	22
Full bunches in padded trays	1967–1975	9
Dehanding/plastic field boxes	1976–1983	8
Field packing	1984–1990	7
Paco pack	1990–1992	3
Cluster pack/mini–wet pack	1993–present	4

Sources: WINBAN and SVBGA files and fieldwork.

[a]Some dates overlap because different systems were used at the same time.

system boxes could have net weights between 26 and 30 pounds, in the cluster pack system the weight must be exactly 34 pounds.

Furthermore, new regulations imposed by the EU created new problems for banana farmers. In particular, the requirement that short bananas had to be packed in separate boxes has led to more quality problems. While the quality of fruit from St. Vincent and the Windwards fell overall in 1995 (see Figure 2.2), a situation that one SVBGA official noted "could only be described as disastrous" (SVBGA 1996b: 2), the problem was most severe in boxes with small bananas, which are more difficult and time consuming to pack than are regular-sized fruit. Thus, while boxes with regular bananas had an average quality level of 73.9 percent, which is generally considered poor and below the Windwards' goal of 80 percent, the percentage for boxes with small fruit was an appalling 56.4 (SVBGA 1996a: 16)—which has helped contribute to low banana prices.[11]

Outrage and frustration best describe the Restin Hill reaction to the forced adoption of cluster packing and the new EU regulations. In 1994, a few grumbled about going on strike but quickly realized that such a move would be futile. Another disgruntled banana grower likened the new system to slavery. The significantly intensified labor burden, coupled with weak banana prices in 1993, 1994, and 1995, led to a reduction or, in some cases, an abandonment of banana growing by some households and a readjustment of labor mobilization strategies in others.

In reviewing such changes from an even longer-term perspective, what is remarkable is not only the numerous innovations that villagers have adopted but also that the pace of technological change has accelerated over time. The general trend is a decline in the length of time during which each system of harvesting and packing was most widely practiced (see Table 5.1). Moreover, changes have occurred within recent systems, such as shifting from divided to dividerless boxes in paco pack in 1991 and the mandatory use of scales and separate packing of small bananas in the cluster pack system in 1995. And such technological changes also have important implications for the social organization of labor in banana production.

The Social Organization of Labor in Banana Production

Most contract-farming enterprises involving peasants are characterized by labor-intensive production, and the Windwards banana industry is no exception. Consequently, it is difficult for one person to manage his or her banana fields alone. Assistance is required from household members, other relatives, friends, and/or wage laborers. Moreover, labor requirements can vary over time not only in response to differences in areas cultivated and age of fields but also in relation to changing technological imperatives, which can necessitate alternations in labor procurement strategies. Thus, growers must creatively organize and manage a sufficient labor force, an endeavor that also clearly differentiates them from wage laborers and calls into question the applicability of the concept of deskilling.

Farmers' involvement in the banana industry is sometimes limited by their labor availability. Indeed, what I found particularly revealing was the almost uniform response of villagers to a question concerning their main constraint in banana production. Given the small areas most farmers control, I clearly expected land shortages to be the primary concern. It was not. Rather, villagers complained largely about their labor constraints, with the common refrain, "Can't get the help." Reflecting the importance of labor constraints, elderly individuals who once produced bananas gradually reduced their holdings in the crop because of their inability to obtain sufficient help and the decline of their own powers to perform the labor-intensive tasks in banana production; they tend to shift to less labor-demanding crops such as plantains and ground provisions. Similarly, many involved in trade work and other forms of off-farm employment will cultivate these other crops, but not bananas, because of their limited time available for agricultural pursuits.

Both male- and female-headed households are actively engaged in cultivating the crop, though the former have more land under bananas, reflecting, in part, their larger holdings in general and greater availability of household labor. Female-headed households with the head as the only adult farmer are at somewhat of a disadvantage in recruiting sufficient labor for banana production, though several such households nevertheless successfully managed their fields, in part by obtaining male help through a variety of social relationships, such as visiting unions or reciprocal assistance. Although both men and women of the same household often have their own individual plots of bananas, they usually work together in cultivating and especially in harvesting them.

Men, women, and children all work in banana-related activities, with men spending the most time on the endeavor. A time allocation study that I conducted in Restin Hill reveals that male heads of households cultivating at least one acre of land spent 20 percent of their total time in banana production, compared with only 10 percent for female heads and wives and common-law spouses of male heads.[12] Many tasks are more commonly performed by men, but both men and women can be found in all aspects of cultivation and harvesting, with reliance on female labor more extensive in female-headed households. Men are more likely to dig banana holes, plant the crop, apply pesticides, plume, sleeve, deflower, and dehand. Women, in contrast, are more likely to pack bananas at harvest time. Both males and females above the age of ten or so often help in carrying or "droghing" banana boxes from field to road. Even younger children sometimes assist with such tasks as carrying empty banana boxes from house to field. It is not uncommon to find children taking off from school to help during harvest days, as is true generally in the Windwards (McAfee 1991: 120).

In Restin Hill, the large majority of labor in banana production comes from the household members themselves, a characteristic they share with other peasantries. The time allocation study that I conducted indicated that 86 percent of all labor inputs into banana production were provided by the households of members controlling the plots, 8 percent by other relatives and friends, and 6 percent by wage laborers. These latter two nonhousehold forms of labor require consideration because their significance for the production process is more substantial than the percentages imply. Specifically, they enable growers to cope with particular, episodic peaks in labor demands that would otherwise constrain production considerably.

Lehmann's (1986: 609) observation about the importance of peasant "access to those parts of the labour market that others cannot reach" is relevant not only to household labor but also to assistance from relatives and friends, which villagers discuss in terms of "helping" and "swap labor." The former refers to labor provided without the expectation of a near-term, comparable reciprocal response, a reflection of close personal ties between the parties involved. In the latter, which is more frequent, those providing assistance expect comparable reciprocation in the near future. Tasks accomplished through swap labor are usually physically demanding, such as digging banana holes, carrying packed banana boxes, or tillage in the case of planting ground provisions, though other activities are sometimes included. Interestingly, 7 percent of all the time that the sample households allocated to banana production was spent in the gardens of relatives and friends while participating in help and swap labor relations—which indicates a close balance with the time others spent assisting them in their own fields under similar social relations.

Swap labor relations are usually on a continuing basis and depend on trust and goodwill. The size of groups involved ranges mostly from two to four, though larger ones can sometimes be found. The basis for such groups varies, with friendship among individuals of roughly similar age and common church group membership frequent components. People note that work seems to go faster when more than one person is involved, as they discuss such varied issues as an upcoming dance, the merits of a recent movie shown on television, the discovery of a new cattle tick in St. Vincent, the comparative dangers of different pesticides, or events concerning the upcoming carnival season.

People also concur that swap labor is less common than in the past, a decline found elsewhere in St. Vincent (see Rubenstein 1987); such a trend is, in fact, widespread among peasants as commodity production in general expands and economic differentiation increases (Grossman 1984a). Villagers suggest that the lower incomes characteristic of the past meant that fewer could hire labor then as compared with the present and thus they relied on swap labor to recruit extra-household assistance. They also complain that people today are less likely to reciprocate adequately for swap labor. As one female grower who no longer preferred to engage in swap labor asserted, "Some don't give you back your days, so I finish with it."[13]

Participation in swap labor varies by age and gender. Men are more involved than women. Men provide most of the labor in banana production, and they tend to do the more physically demanding tasks, which are more likely to be included in swap labor relationships. Younger men, in particular, participate more frequently in swap labor relations in banana production than do older males. Their limited financial resources to hire labor and the generally smaller size of their children constrain their ability to mobilize sufficient labor through alternative means. In contrast, older men, particularly those with several acres of land in bananas, prefer to mobilize their own larger household labor forces or hire workers when necessary. Some do not want to be obligated to others and feel it is beneath their standing to have to "work" for someone else, even if it is a swap labor arrangement. Jealousy in the village is also relevant, as older men usually have larger landholdings than others. One of the richest men in the village declared, "Once you come up a bit, others don't want to help. Feel you are a bigshot. Some don't want to see I have more than he."

Wage labor is also a significant dimension in banana production for some households. The extent of hiring depends on both areas planted and family size.

Those with larger areas are more likely to hire wage labor, but exceptions are evident. For example, one household with 4.7 acres of bananas never hired labor, relying instead on husband, wife, and four hard-working, resident children aged sixteen to twenty-one. In contrast, households with smaller holdings but fewer family members may resort to hiring labor occasionally, such as for droghing banana boxes or digging banana holes. Only two households in the village (neither was in the sample), one with over ten acres of agricultural land and the other with over four, employed wage labor regularly on a daily basis. As an indication of the comparatively limited use of wage labor in Restin Hill compared with that in other, nearby villages, the five members of the sample who engaged in agricultural wage labor on a regular basis all worked for people living outside the village. And no household in Restin Hill relied primarily on wage labor in their banana fields.[14]

In the late 1980s, growers complained about a shortage of wage laborers. This scarcity reflected, in part, the growing need for labor, as the intensity of banana production associated with field packing increased and the acreage in bananas expanded attendant with the rise in producer prices. The shortage of labor affects growers' relations with wage laborers. Many farmers attempt to maintain good relations with their workers, with some having relationships lasting many years. Indeed, many feel at the mercy of their wage laborers and are hesitant to chastise them, fearing they will lose access to a valuable and scarce resource and obtain a bad reputation among other workers. Abusive practices by agricultural employers—common in many developing countries—are rare in the village. Indeed, growers add certain benefits to maintain their good relations with their employees, such as occasionally allowing employees to harvest provisions from their fields or providing them with lunch.

At the same time, many farmers have negative views of agricultural workers in general. They complain that they do not receive adequate work for the wages provided, feeling that they obtain half a day's work for a full day's wage. Most can recount instances in which they had made arrangements with wage laborers to show up on a particular day, only to find that the individual failed to arrive as promised. A more extreme view was represented by a sign posted by one banana grower who often hired laborers: "Beware of shooting anything on this place that isn't moving. It's probably the hired help." The pressure farmers feel in relation to workers stems not only from the shortage of labor but also from the regimentation of the harvesting regime. Bananas must be harvested at a certain stage of development, when approximately three-quarters full, to be able to remain unripe until reaching the United Kingdom, the exact grade harvested depending on the season. If harvesting is delayed until the following week because of insufficient labor, the fruit may become too full and commercially unacceptable. Thus, workers failing to show up on harvest days presents a physical and possibly a financial hardship for growers.

Also contributing to labor scarcity is the general reluctance to work for others in the village. Not only are people hesitant to make others rich by their own labor, but their own friends may also discourage them from engaging in wage labor, chiding them for working too hard. As one villager noted, explaining the hesitancy of people to be wage laborers, "They don't want to feel anyone better than he." Undoubtedly, the reservation concerning working as agricultural wage laborers stems in part from the time of slavery and the exploitation of workers on estates in the postemancipation era. But another dimension is also important. The desperate poverty in many developing countries that drives people into exploitative wage labor relationships is absent in the village and most other banana-producing areas. Because certain agricultural products are abundant during particular seasons, people can often obtain free access to green bananas, breadfruit, coconuts, mangoes, and other fruits. Many adult children live with their parents, which limits their cash income requirements. Remittances from overseas spouses, boyfriends, and relatives also lessen the need to work for others (Rubenstein 1983).

Growers hire labor for a variety of tasks, including digging banana holes, pluming fields, and packing and carrying banana boxes during harvest days. But for tasks that are not physically demanding, that require considerable care, or that do not have rigid scheduling requirements, farmers prefer to use their own household labor. Thus it is rare for growers to hire for deflowering or sleeving. Similarly, they hesitate to employ wage labor to apply fertilizers and pesticides, fearing possible careless application and waste of expensive agrochemicals.

On banana-harvesting days, agricultural employment is at its highest levels. When hiring packers, growers prefer women because they are believed to be more skilled and careful packers than men. Interestingly, the preference for females because of their alleged greater care and dexterity echoes a common refrain of employers in many industries, especially in electronics assembly, in the NICs; but in contrast to the situation in such areas, where women are paid less than men, in St. Vincent female banana packers receive the highest wages of all banana employees, given the crucial importance of the task. Packers usually receive between EC$25 to EC$45 per day, depending on the amounts harvested.

The most frequent need for wage labor is in carrying banana boxes from fields to roads. Those hired include adult men and women and children as young as ten. Payments varied. In the late 1980s, it was usually EC$20 per day for men,

EC$15 to EC$20 for women, and somewhat less for children. The premium for men "droghers" reflected their tendency to carry two to three boxes per trip—whereas women normally transported between one and two—and growers' beliefs that men carried banana boxes faster than women did.

Utilization of these different forms of labor procurement—from household members, from relatives and friends in the form of helping and swap labor, and from wage laborers—is affected, in part, by farmers' wealth, household size and composition, age, gender, and interpersonal relationships. At the same time, technological change mandated by capital and by the state and fluctuations in banana prices also affect labor procurement strategies, revealing that they are highly dynamic through time. The manner in which growers responded to the introduction of cluster packing and declining banana prices in 1993, 1994, and 1995 is illustrative.

These two changes created a squeeze on farmers, as cluster packing intensified their labor needs at the same time that banana prices—and hence their incomes—fell. Responses varied considerably. Most farmers reduced their financial outlays, including purchasing and applying fewer agrochemical inputs. A decline in the level of field care, such as spending less time on weed control and pluming, was also very evident. Some, especially older banana growers or those with limited available household labor, simply ceased commercial banana production altogether, relying increasingly on the production and sale of less labor-intensive food crops. Others reduced their areas planted in bananas to what household members alone could manage, thus obviating the need for wage labor. Indeed, a visual inspection of the village in 1995 revealed that the amount of land planted in bananas was roughly 20 percent less than that in 1989, reflecting similar declines throughout St. Vincent and elsewhere in the Windward Islands. Part of the hesitancy about relying on hired labor, besides the decline in banana prices, reflected the increased cost of wages, as droghers were now demanding EC$25 to EC$30 per day in 1994 and 1995. Both the effort and the skill content of their work had also increased as a result of cluster packing; previously, they had only had to carry boxed bananas from field to road in the field pack and paco pack systems, but now, in addition, they also had to shred green banana leaves to provide cushioning for the freshly cut clusters, carefully arrange them in the trays, and then carry them to the central packing areas. Another response has been to rely more on swap labor. Farmers—both men and women—who previously shunned the relationship became dependent on it. Those who used to harvest on a biweekly basis now resorted to weekly harvesting, working in their fields one week with the help of swap labor and returning the assistance to their partners the next. Nonetheless, many who increased reliance on swap labor still had to hire workers as well. Those with larger holdings have had to hire extra people just to dip and wash the fruit, a task that now receives EC$20 per day.

Such changing patterns of labor mobilization are inherent in the dynamic nature of contract farming. They are a fundamental element of conception in the production process that is inadequately captured by the concept of deskilling. Moreover, their significance clearly distinguishes such farmers from wage laborers, making the concept of "disguised wage labor" inapplicable to the complexities of the labor process in banana contract farming.

At the same time, the need to manage their labor forces creatively reflects adjustments to the power exercised by capital and the state in the harvesting/packing stage. And such control has also led to technological changes requiring increasing skills in banana contract farming over time. But capital and the state have been unable to regulate the phases of cultivation effectively because of both environmental and political-economic conditions. Such influences—which are at the heart of political ecology—are also significant in the analysis of the relation of banana export production to local food production and food import dependency.

Six

The Food Question

When landing at the Vincentian airport on the outskirts of Kingstown, arriving passengers are greeted by a colorful poster displaying the message, "Produce More, Import Less." This seemingly simple phrase in reality reflects a profound and complex problem plaguing St. Vincent and the rest of the English-speaking Caribbean—declining local food production and burgeoning food imports (Long 1982; Brierley and Rubenstein 1988; Thomas 1988; Richardson 1992). Where once locally produced ground provisions, breadfruit (*Artocarpus communis*), bananas (*Musa* AAA Group), plantains (*Musa* AAB Group), and meats were the mainstay of Vincentian diets, now people depend more on imported flour, rice, and frozen chicken. The key issue here is the role played by contract farming and intervention by capital and the state in the production process in relation to these transformations.

Reflecting the tendency in the broader development literature to blame export agriculture in general for such problems, researchers in the Caribbean claim that the banana industry in St. Vincent and elsewhere in the Windward Islands has been responsible for undermining local food production and increasing dependence on food imports (see Ellis 1975; Marie 1979; Long 1982; Rojas 1984; Axline 1986; Thomson 1987; Thomas 1988; Nurse and Sandiford 1995). For example, Rojas (1984: 45) declares:

> The special characteristics of banana production meant medium and small-sized farms were easily incorporated into the production of export crops. This has made export production even more competitive for land, labour and capital with general food production than it is normally in plantation economies. Food production for the local markets is to date a markedly residual activity in the Eastern Caribbean resulting in persistent problems of food supply and the pronounced enclave character of export production.

The problem with such analyses is that they fail to specify the precise means by which banana contract farming is supposed to undermine local food production. A basic deficiency in this literature is the absence of detailed, village-level case studies examining the changing role of production in agricultural systems. This study reveals that the relationship among banana exports, local food production, and food imports—the "food question"—is much more complex than the literature suggests.

The discussion begins with a brief overview of the nature and extent of food import dependency. It then reveals some of the intricacies of the local food production system, which are essential for understanding the relation between food production and banana contract farming. I also examine systems of marketing and their dynamics to illuminate the crucial problems that farmers face when attempting to sell their local food crops. The analysis shifts to an evaluation of possible conflicts in the productive sphere generated by the control exercised by capital and the state over the peasant labor process. I conclude with a discussion of the significance of the differing political-economic contexts of production and marketing for understanding the "food question."

Food Import Dependence

Food importation is certainly not new in the English-speaking Caribbean, as reflected in Stephen's (1830: 261) concept of "foreign-fed colonies" during the era of slavery. Being a "home-fed colony" during slavery and subsequently having an expanding peasantry, St. Vincent has historically been less dependent on food imports than many of its neighbors; thus in his study of agriculture in the region, Engledow (1945: 216) noted that in 1937, the value of food imports per person in St. Vincent was low compared with the situation in many of the other British Caribbean colonies.[1] Nonetheless, reliance on food imports grew slowly in the first half of the twentieth century. But since World War II its growth has been especially marked, with a particularly significant upswing in the 1980s (Figure 6.1). Although food imports have certainly grown, interestingly, the percentage of all imports composed of food products has not, having fluctuated mostly between 20 and 30 percent.[2] In essence, the growth of food imports has paralleled the general rise in overall imports during this period, all of which have contributed to a worsening visible trade imbalance (see Figure 3.1).

The United States is clearly the major source of food imports, supplying 43 percent of the total in 1992 (SVG Statistical Unit files). The United Kingdom and Trinidad were the second and third most important sources, supplying 16 and 11 percent, respectively. Although these three countries provided 70 percent of all food imports, St. Vincent's food network is surprisingly global for such a small country, it having imported foods from fifty-four different countries from every region in the world in 1992.[3] Thus, in a single day or two, a Vincentian might have for the morning meal tea from Malawi, Kenya, or China,

FIGURE 6.1. Food Imports into St. Vincent, 1956–1993

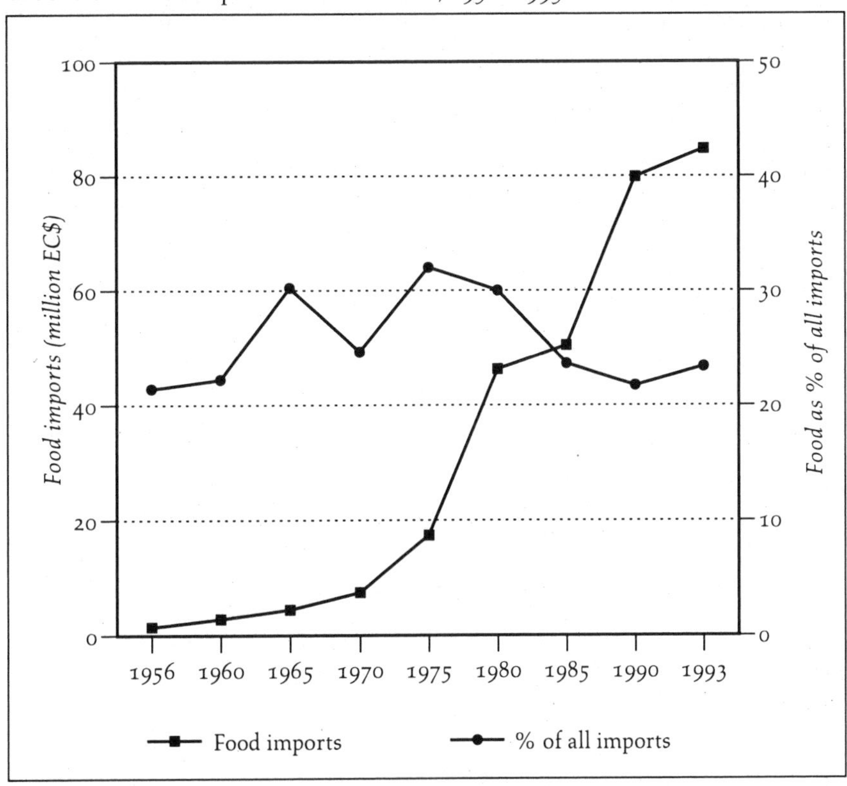

Source: SVG Statistical Unit digests of statistics and files.

perhaps mixed with some powdered milk from Holland or France, along with bread made from flour imported from the United States and coated with margarine from Canada. A later snack may include biscuits from Jamaica or Trinidad and canned sausage from Venezuela or Brazil. If really hungry, a large lunch follows, possibly including rice from Guyana and kidney beans from Belgium, perhaps washed down with orange juice originating from Belize. The evening repast often includes frozen chicken from the United States, likely seasoned with garlic from Mexico or Albania and a variety of spices from Singapore. Certainly, some locally produced foods will also grace Vincentian tables, but the importance and global nature of the food system is evident.

Food import dependence is not just an urban phenomenon in St. Vincent. Even though Restin Hill is primarily a farming community, most of the food it consumes is also purchased—and most foods are imported—a pattern that is widespread in rural areas throughout most of the island. Given the unequal distribution of landholdings in the village, the pattern is not surprising. For many Restin Hill households, agriculture does not provide sufficient support or sustenance for their members, a situation that forces such households to engage in wage labor and to purchase food. But even most households that are dependent primarily on agriculture also buy much of their food.

An examination of the frequency of consumption of foods providing the major sources of carbohydrates and animal protein in the village is illustrative.[4] The three most frequently consumed major sources of carbohydrates are rice and flour, which are imported, and the ground provision dasheen (*Colocasia esculenta* var. *esculenta*), grown by the villagers. Flour is eaten more frequently than any other food, appearing in a variety of forms, including bread, cakes, dumplings, and "bakes" (which are prepared by frying flour mixed with baking soda, sugar, and salt). Rice is usually a component of mixed dishes, such as rice and legumes and also "pelau," which normally includes rice, legumes (local and imported), and chicken. Villagers cook dasheen in a variety of ways; a particularly popular form is the "boilin," which consists of dasheen and other provisions, green bananas, dumplings, plantains, fish or meat, and, if in season, breadfruit, all of which are boiled in water and oil. The main animal protein sources, referred to as "flesh," are eggs (from the village and locally produced ones sold in Kingstown), local fresh fish, and imported chicken, especially chicken back and wing. Villagers consume eggs mostly at "tea" (breakfast), and fresh fish purchased in the Kingstown fish market is usually part of weekend meals in "boilins." Chicken is the most frequently consumed source of animal protein. With respect to these foods, as well as other sources of animal protein and carbohydrates, the majority—60 percent of the foods providing the main sources of carbohydrates and 67 percent of those providing the main sources of animal protein—were imported.

The literature on the globalization of agriculture suggests that we cannot understand food patterns in developing countries without considering the influence of food systems in advanced capitalist countries, a point that is certainly relevant to the case of Restin Hill and the rest of St. Vincent. In particular, consumer patterns in the United States have influenced the growth of chicken imports, particularly that of chicken back and neck, which made up 41 percent of all chicken imports by weight in 1994 (SVG Statistical Unit files).[5] The fast-food chicken industry in the United States, composed of such outlets as Kentucky Fried Chicken, McDonald's, Chick-fil-A, Hardee's, Bojangel's, and, most recently, Wendy's, has expanded dramatically, creating an insatiable demand for particular chicken parts—wings, breasts, and legs. The problem of marketing the other, lower-value parts of the chicken, the neck and back, has been solved by U.S. agribusiness by exporting them to areas such as the Caribbean, being able to sell them at low cost because of the high volumes of chicken produced for

FIGURE 6.2. Poultry Imports into St. Vincent, 1970–1994

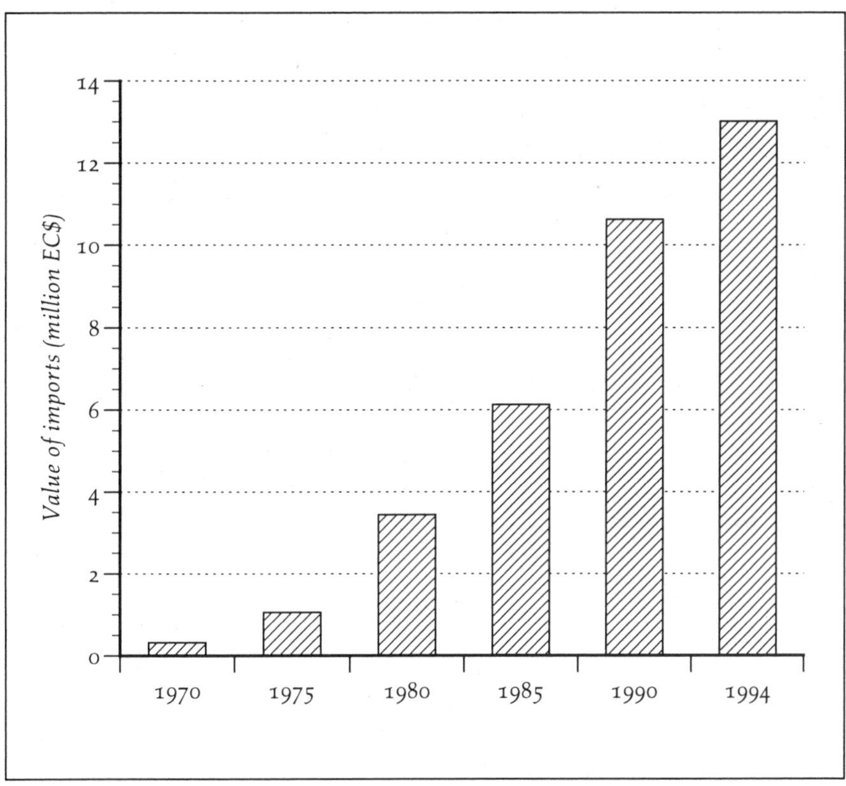

Source: SVG Statistical Unit files.

the U.S. market. Indeed, the growth of frozen chicken imports since the 1970s has been meteoric (Figure 6.2).[6]

Certain explanations concerning the growth of food import dependence in the Caribbean generally are not particularly relevant to the case of contemporary St. Vincent. Some researchers (see Long 1982: 768; Mintz 1983: 10) have argued that people in the English-speaking Caribbean prefer imported foods because of prestige considerations, as consumption of local foods is felt to be associated with low status and a bitter past. Such an explanation would have been more applicable in the past, when use of imported foods was less extensive, but today their consumption is so widespread that prestige considerations would be only a minor element of the explanation.[7] Also, linking food import dependence to both tourism and the penetration of fast-food franchises from the United States, such as McDonald's, Burger King, and Kentucky Fried Chicken (McElroy and Albuquerque 1990; Richardson 1992: 109, 111; Thomas 1988), is of limited relevance. In St. Vincent, tourism has not grown as it has in

FIGURE 6.3. Banana Exports and Food Imports, St. Vincent, 1970–1994

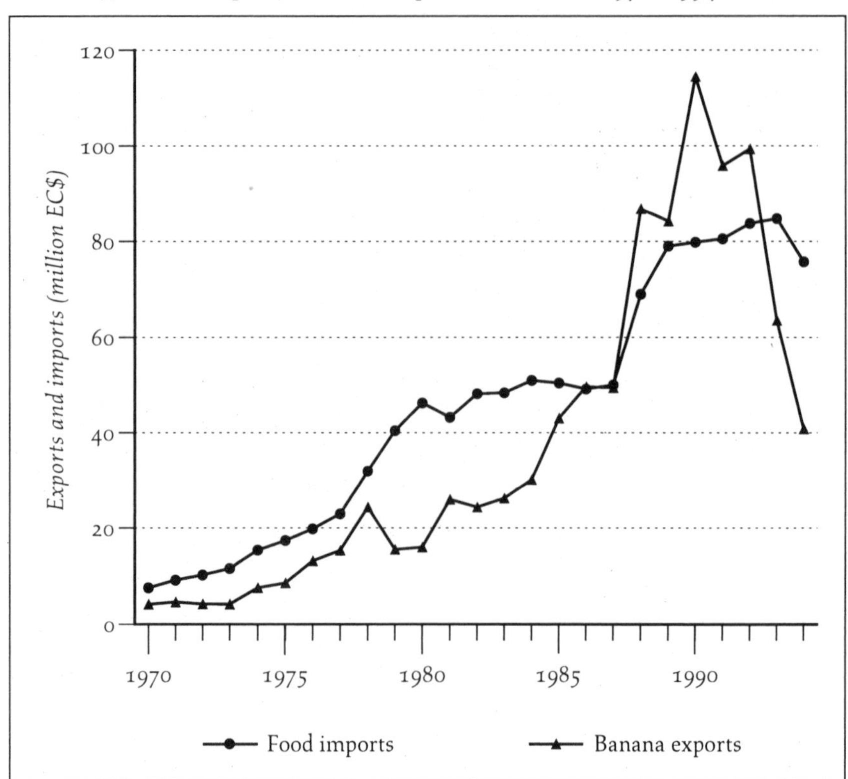

Sources: Grossman 1993; SVBGA annual reports; SVG Statistical Unit digests of statistics and files.

nearby islands, and the country has only one fast-food outlet, a Kentucky Fried Chicken restaurant in Kingstown.

The current tendency in the literature to blame the banana industry for rising food imports provides only a superficial explanation for a rather complex situation. A careful inspection of changes in indicators over time suggests that the alleged close relationship between food import dependency and export banana contract farming is certainly suspect (see Figure 6.3). From 1970 to 1978, a close correspondence between the two is evident, but food imports continued to rise from 1979 onward, whereas banana export incomes started to fall. A close relation between the two resumed only for the period 1986–1988. While food imports maintained a very slow but steady increase from 1988 to 1993, banana export incomes exhibited wild fluctuations from year to year that have limited relation to the pattern of food imports. Only in 1994, when banana incomes fell a precipitous 36 percent, did food imports move again in tandem, although to a much smaller degree, only 11 percent.

While an intimate link between export banana production and food import dependence is not supportable, banana incomes do have certain influences that need to be considered. Understanding the role of banana production requires disaggregation of two levels—national and local. At the national level, the rise of banana export incomes up to 1992 had spurred growth of the Vincentian economy, stimulating transportation, construction, and retail activities, and thus contributed to wage income growth in both rural and urban areas. At the village level, increasing banana production itself led to more agricultural wage labor opportunities. Certainly, a portion of such wage incomes that grew at the national and village levels was spent on imported foods. For those who grew bananas and food crops, however, the linkage is surprisingly less evident, owing to the nature of household expenditure patterns. The general tendency is for women to use their incomes from the sale of local food crops to buy dietary items for their households and use banana incomes for that purpose only if revenues from food crop sales are insufficient. In contrast, villagers generally reserve banana incomes for such lump sum expenditures as school fees, clothing, housing, and land purchases.

Local Food Production in Restin Hill

Restin Hill is located in the sheltered Marriaqua Valley, one of the most extensively cultivated areas on the island and often referred to as the breadbasket of St. Vincent. Because farmers in Restin Hill work gardens situated at between 680 feet and 1,750 feet in elevation, many of their plots are higher than most others in the valley; thus the seasonal impact of the dry season on food production is less severe for these farmers than for most other villagers in the valley and elsewhere in St. Vincent.

The village itself has a long tradition of diversified food production, cultivating a variety of ground provisions, bananas, plantains, fruits, and vegetables for home consumption and for sale in local food markets.[8] As elsewhere in the Caribbean, the complex of crops one finds today has a rich and varied past; some were first grown by the Amerindians, the later Europeans introduced others, and a number of important crops also came with the African slave trade—all of which have been complemented by additions from Asia and Oceania (Mintz 1985; Berleant-Schiller and Pulsipher 1986).

Food crops are subsidiary to bananas, but their relative importance varies according to the type of household (Table 4.1). They covered 29 percent of the land under crops, less than the 77 percent devoted to bananas.[9] After bananas, dasheen is the most important crop in the village. In addition to providing an edible corm, which is a dietary staple, it also yields leaves that are the basis of the nutritious and much-relished green callaloo soup. Other ground provisions found frequently in Restin Hill gardens include eddoes (*Colocasia esculenta* var. *antiquorum*), tannia (*Xanthosoma sagittifolium*), yams (*Dioscorea* sp.), and sweet potatoes (*Ipomoea batatas*).[10] Plantains, a relative of the banana, are also in abundance, along with a variety of traditional bananas, consumed ripe and cooked. What people call "greens" are also prominent, especially tomatoes (*Lycopersicon esculentum*) and cabbage (*Brassica oleracea*), but also lettuce (*Lactuca sativa*), cucumber (*Cucumis sativus*), carrots (*Daucus carota*), peppers (*Capsicum* spp.), a variety of beans (*Phaseolus* sp.), and pigeon peas (*Cajanus cajan*). These food crops have different maturation lengths that help stagger the workload in gardens (Table 6.1). In addition to such garden crops, villagers also plant a wide variety of trees, including breadfruit, cinnamon (*Cinnamomum zeylanicum*), coconuts (*Cocos nucifera*), orange (*Citrus sinensis*), plumrose (*Syzygium jambos*), golden apple (*Passiflora laurifolia*), grapefruit (*Citrus paradisi*), avocado (*Persea americana*), mango (*Mangifera indica*), nutmeg (*Myristica fragrans*), and soursop (*Annona muricata*). Breadfruit, descended from saplings brought to the island during Captain Bligh's historic voyages, is a much-relished, starchy fruit that is prized in "boilins" and is also enjoyed roasted. The ubiquitous coconut tree provides a ready thirst quencher for the hard-working villagers. And people consume fruit not only whole, a delicacy that children especially seem to like, but also in the numerous sweetened fruit drinks that villagers prepare to accompany their meals throughout the year, the exact variety depending on the season. Indicating the wide diversity of agricultural commodities, the villagers in the sample grew thirty-nine different garden crops, not including exportable bananas, and thirty-two different tree crops, with most gardens containing between ten and twenty garden and tree crops.

Food crops are found in a variety of contexts—covering an entire plot, being intercropped with bananas, or being grown separately while bananas are cultivated on another part of the same field. It is rare to find extensive areas covered by local food crops in a single garden, the largest in the Restin Hill sample being only 1.2 acres, whereas the comparable figure for bananas is 4.3 acres. In the case of ground provisions, market conditions do not warrant cultivating larger areas. And farmers hesitate to plant garden greens more extensively because of the high market and environmental risks; prices fluctuate considerably, and pests and rainfall can destroy entire crops.

The work burden in local food crop production is distributed throughout the

TABLE 6.1. Length of Time to Harvest Major Food Crops

Crop	Time to Harvest
Beans	6 weeks
Cucumber	2 months
Tomatoes	3 months
Carrot	3 months
Cabbage	3 months
Corn	3 months
Sweet potato	5 months
Eddoes	5–6 months
Yam	7 months
Dasheen	8–9 months
Tannia	10–11 months

Source: Fieldwork.

year because the crops that villagers cultivate have different seasonal requirements and the impact of the dry season is moderated somewhat by the altitudinal range exploited. Certain crops, such as beans and carrots, are planted during most of the year. Many villagers believe that it is best to plant dasheen from May to November to take advantage of the rains, but others still plant the crop in drier months at higher elevations. Complementing the seasonality of dasheen is that of yams, cabbage, and tomatoes, which most villagers plant from December to March. Cabbage and tomatoes, in particular, do poorly in the rainy season, as intense precipitation can wash away planted seeds, knock blossoms off tomatoes, encourage fungal diseases, and spoil cabbage plants.[11]

The Pattern of Food Crop Production

Discussion of the pattern of food crop production is relevant to understanding the role of banana contract farming in several ways. First, it indicates that banana production is technologically much more complex, reflecting the qualitatively different nature of contract farming. Second, it helps illuminate the manner in which bananas and food crops are integrated into the agricultural system, which is crucial for understanding the food question. Third, it provides a window through which to view the popularity of banana production. And, fourth, it is an arena in which peasants have considerable autonomy, because capital and the state do not intervene to any significant extent in the process of local crop cultivation.

Land preparation—soil tillage—is an adaptation to the steep slopes on which peasants cultivate, a reflection of the heritage of the island's political economy. It is the most physically arduous aspect of all agricultural activities and, at the same time, exhibits considerable diversity in practices. Villagers identify three systems of tillage: "range and cover," "range and fork," and "banks," each particularly suited to different crop combinations. In both "range and cover" and "range and fork," farmers use a hoe to prepare ridges along the contours of slopes to reduce the threat of erosion and to drain water off to the sides of gardens. In range and fork, farmers dig or "punch" holes with a fork in the furrows between each ridge and usually plant dasheen in them, while cultivating such crops as beans, cucumbers, and carrots along the tops of ridges. In range and cover, in contrast, farmers prepare somewhat larger and broader ridges and plant only atop them, not in the furrows, a system most suited to such crops as sweet potatoes, cabbage, tomatoes, carrots, and eddoes. Farmers also dig "banks," or mounds, which are less common than the other two systems in Restin Hill, a form of tillage reserved primarily for growing yams and, to a lesser extent, eddoes, with the flat spaces among the mounds allocated to the cultivation of dasheen and other crops. As in the case of the types of crops grown, patterns of tillage have a complex heritage, with roots in pre-Columbian Amerindian practices, sugar cane cultivation methods used in estates during the era of slavery, and traditional techniques employed in West African gardens (Berleant-Schiller and Pulsipher 1986).

The exact sequence of subsequent events varies according to the method of tillage and crops cultivated, but a general outline is possible. Fertilization occurs once or twice, weed control twice, and molding once. Villagers use synthetic fertilizers widely, obtained mostly from the SVBGA. Some also apply manure from cattle kept adjacent to their fields, especially when planting yams. Many use a variety of pesticides, especially in the cultivation of "greens." Weed control is more difficult and time consuming in food crop production than in banana cultivation because food crops can be killed by contact with the herbicide Gramoxone.[12] Thus, villagers often hand-weed their food crop fields, either by pulling out the unwanted vegetation by hand or by cutting it, with men using cutlasses and women employing hoes for the task. When they do use Gramoxone in such fields, it is primarily for spraying along the backs of ridges where no food crops are planted.

The final stage is harvesting. In contrast to bananas, which must be harvested in a fairly rigid, timely manner, food crops can be left in the ground for varying time periods after they are ready to harvest, providing a certain level of flexibility (Table 6.2).

After the harvest, farmers fallow their fields for varying periods, most of which are shorter than one year and in many cases just a few months. Fallows

TABLE 6.2. Length of Time Provisions Can Be Left in Ground after They Are Ready for Harvest

Provision	Time Provision Can Be Left in Ground
Sweet potato	1 month
Yam	1–1½ months
Eddoes	1 month
Dasheen	1 month
Tannia	2 months or longer

Source: Fieldwork.

were longer in the past, but the introduction of synthetic fertilizers has enabled villagers to reduce them. The expansion of acreage under bananas, particularly that which occurred in the boom of the 1980s, has also put pressure on fallow periods. And, of course, the short fallow periods also reflect limited land availability for most households.

A discussion of just a few of the possible combinations and permutations of crop successions helps illuminate the varied ways in which bananas and food crops are integrated. In many cases, bananas and food crops are intercropped on a regular basis; in such a pattern, most of the food crops are harvested within nine months, while the bananas last for two to three years, after which the field is interplanted again.[13] Others vary the combinations over time. For example, in one field a villager first grew bananas, then used the plot for a cattle pasture, subsequently grew yam and dasheen there, and afterward intercropped bananas with dasheen and "greens." In another case, a farmer first planted bananas alone, then grew only tomatoes and cabbage there, and once again returned to banana cultivation. In contrast, some regularly replant the same field or portion of a field in food crops alone without bananas, though often varying the food crop mix over time. Villagers also employ food crop cultivation to actually enhance banana yields. They explain that when bananas begin to yield poorly, often after continually being replanted alone in the same field for many years, it is helpful to plant only food crops there before replanting another banana crop. The extensive tillage associated with food crop cultivation is believed to improve soil structure and aeration, which enhances subsequent banana yields. Several farmers assert that the preparation of banks, in particular, is helpful in this regard.

Also important are cattle, which people raise in one of two ways. In the "pen patch" system, they tether their animals in a small, flat area called a pen patch. Farmers generally plant elephant grass (*Pennisetum purpureum*) surrounding their pen patches and often around the boundaries of nearby gardens to serve as sources of fodder. In this labor-intensive system, villagers feed their animals the elephant grass, along with weeds cut from the edges of paths and from nearby gardens, twice each day. Traditionally, villagers gathered manure accumulated at the pen patches, carried the material in baskets, and distributed it in their fields as fertilizer. Farmers occasionally still use such manure today, though much less frequently than in the past. In the second system, called "tie-out," people tether their cattle to a long rope or chain on fallow land, moving the animal once each day to a new part of the field. Tie-out is less labor intensive than the pen patch system, as farmers do not have to cut grass for their animals, but it is feasible only for those with sufficient amounts of land. In addition to distributing manure in fallowed fields, cattle grazing on such land has another crucial benefit—restoring soil structure. Villagers explain that repeated cultivation makes the soil "loose," which reduces yields. The basis of the problem is the nature of Vincentian soils, which are light in texture compared with those in the other Windward Islands. While fallowing alone helps to "harden" the soil, which is crucial for adequate yields, the trampling of cattle improves the process considerably, helping to bind the soil together—or, as one farmer described it, to "get the land back in order." Such trampling likely increases soil bulk densities, which could reduce loss of soil moisture through evaporation and percolation (a particular benefit in the dry season) and provide a better soil environment for the rooting and sprouting of plants, because the roots could establish better contact with soil surfaces and absorb nutrients more directly (Art Limbird, pers. comm., 7 December 1995).

As in banana production, all household members can help in food crop production, and the division of labor by gender is not rigid. Similarly, some tasks are more likely to be performed by men and some by women, though both can be found involved in all aspects of local food production. Soil tillage, which is the most physically demanding aspect of food crop cultivation, is generally the province of men and the task for which people most frequently hire labor in food crop cultivation. Weeding and molding are more closely associated with women, with weeding another task sometimes allocated to wage laborers. Except in the case of yams, both men and women are actively engaged in planting and harvesting, with men more involved in yam cultivation. Pesticide application and cattle raising are usually, though not exclusively, male activities.

Food Marketing

In the literature on peasantries in developing countries, households are generally portrayed as producing food primarily for home consumption and selling the "surplus" (see Guillet 1981). The situation in Restin Hill and much of the rest of St. Vincent is dramatically different. Although villagers here produce local food crops for both home consumption and for sale, the major emphasis for most farming households is clearly on income earning and secondarily on subsistence (Grossman 1993). They estimate that, except for tree fruits and some minor crops, they sell the majority of their food crops. A recent analysis of Vincentian agriculture (Wedderburn 1995: 39) reveals that the Restin Hill pattern is not unique; it notes that "Vincentian farmers have a high degree of market orientation. With few exceptions a major proportion of crop production is sold. There is no evidence of principally subsistence production." Indeed, in Restin Hill a significant amount of the local food consumed is, in fact, that which could not be sold. Thus, market conditions are a major determinant of the area cultivated in food crops, a pattern found elsewhere in the English-speaking Caribbean (see Moberg 1991). As a result, the better the market is for a crop, the larger the area that will be devoted to it and the more likely it will be consumed more frequently by peasant households. For peasants in banana-producing areas like Restin Hill, however, markets for food crops did not provide the necessary incentive to expand production in the 1980s and early 1990s, though the situation had changed by the mid-1990s.

The analysis of marketing problems requires consideration of the destinations involved. Vincentians sell their crops mainly through two outlets—local markets, the primary one being in Kingstown, and traditional regional markets dominated by female exporters called "traffickers." Several other outlets are also employed, including the St. Vincent Marketing Corporation, a statutory corporation, but such outlets are of minor significance compared with the Kingstown market and traffickers. The two major markets share a common feature—they are both characterized by open markets and all the problems associated with them.

The agricultural abundance of the countryside meets the Vincentian consumer in the vibrant Kingstown market, which, as is typical in the region, combines both retail and wholesale functions. Here, people can purchase a wide range of agricultural products—ground provisions, a variety of bananas and plantains, fruits, "greens," breadfruit, coconuts, spices, and eggs, as well as an assortment of other items, such as handmade brooms, charcoal, cake, bush teas, and chicken feed.

Over the last several decades, the number of sellers and the amounts sold in the Kingstown market have grown. Access to motorized transport has increased both accessibility to the market for people throughout the island and the amounts that each seller can carry to town. Moreover, the Vincentian urban population—the primary customers of the Kingstown market—has also expanded.

The Kingstown market operates Monday through Saturday, with Fridays and especially Saturdays being the busiest.[14] Sellers, who are mostly women, begin to arrive as early as 4:45 A.M., and subsequently a steady stream of vans and pickup trucks carry villagers and their produce to market, with activity finally tapering off late in the afternoon. Sellers can easily be identified by their aprons, in which they keep the money received from marketing transactions and which help protect their clothes from the dirt and dust. As crowds walk by inspecting produce, the air is occasionally punctuated by calls to customers, with such pleadings as, "One dollar here—sweet, sweet lovely oranges here," "Psst, want to get some tomatoes," and "Celery here."

The Kingstown market has two types of sellers. "Hucksters," traditionally located inside the large, galvanized-roofed market building, obtain their supplies by buying from farmers throughout the island and sell each day the market is open.[15] In contrast, "market vendors," who are active primarily on Fridays and Saturdays, traditionally sell just outside the market building. They lay their produce out on roughly built wooden tables or on the ground on mats cut from empty fertilizer bags, with many relying on umbrellas to shield themselves and their commodities from the baking sun. Market vendors obtain their commodities primarily from their own gardens, though some supplement their supplies by buying from other farmers, a practice that people refer to as to "buy and sell over." A typical Restin Hill vendor tends to sell between five and ten different crops on a single day, including dasheen and other ground provisions, plantains, green and ripe bananas unsuitable for export, fruits, and a variety of "greens," especially tomatoes and cabbage. When vendors have more crops from their own gardens than they can market themselves, they will sell the surplus to hucksters.

The pricing structure in food marketing is much more variable and fluid than that in the banana industry, as each seller can set her own prices. Items are sold in a variety of ways. One method is offering items by the piece, as are "water nuts" or green coconuts, which cost about EC$1.00 each, and ripe bananas, prices for which range from one finger for EC$0.10 to two for EC$0.25. Others, such as tomatoes and cabbage, are often sold by the pound. Provisions are normally offered in "heaps," or piles, available in two sizes, costing EC$2.00 and EC$4.00. Hucksters generally charge slightly higher prices than market vendors, as they have to purchase all the items that they sell.

A seasonal dimension to pricing exists. Demand and prices for greens are particularly variable over time, being highest at times associated with holidays—just before Christmas and around June and July, when Carnival is held. Environmental variables are also influential. Vincentian farmers prefer not to plant greens during intense rains, which reduces plantings in September and

October. Planting is also constrained in many areas of the island during the dry season in March and April. As greens take approximately 2½ to 3 months to mature, such a planting pattern reduces supplies at times of maximum demand, contributing to higher prices. The cost of provisions also varies over time, reflecting more complex influences: reduced plantings in the dry season in much of the island and the fluctuating fortunes of export markets, which have longer-term, nonseasonal effects.

Although the social dimension of marketing is important for women, income earning is clearly their paramount concern. Weekly receipts in 1988–89 averaged between EC$50 and EC$150 per marketer, the differences reflecting primarily amounts of land controlled. Returns can also vary weekly, with highest sales usually at the end of the month, when government employees and management in the private sector receive their salaries. But while profitability is important in understanding marketing, it is not the only consideration. Even when prices are low, women continue to participate in the market because it gives them a source of income over which they have control (Mintz 1983). Indeed, they generally do not tell their husbands or common-law spouses how much they earn from their market sales.

As significant as food marketing is for women, a variety of constraints limit food production for the domestic market. Marked volatility in prices can be a disincentive to production. The sale price for tomatoes, for example, can vary seasonally from EC$0.50 to EC$7.00 per pound. While some price fluctuations for such "greens" are fairly predictable, others, particularly for provisions and plantains, are more difficult to forecast. Farmers complain about a recurring pattern that plagues them: When prices for plantains and provisions are high, farmers throughout St. Vincent tend to expand the areas planted in these crops. But the increased production stimulated by the high prices usually leads to excess production in relation to demand, which then drives prices down by harvest time. Expected high financial returns thus often fail to materialize.

Even more important is limited consumer demand. The mostly urban and semiurban consumers prefer imported foods, such as flour and rice, over locally produced crops for several reasons. The issue of cost is paramount not only for urban consumers but also for the rural population that purchases much of its diet. People assert that it is less expensive to feed a family with imported foods than it is with local foods purchased in the Kingstown market (compare Mintz 1983:10). A comparison in relation to the cost of calories and protein of two local and two imported dietary staples confirms their assertions (Table 6.3). Moreover, the rate of inflation in a variety of imported foods since 1964 has been considerably less than that for several locally produced food items (Figure 6.4). These cost differences reflect, in part, contrasts in the technical and politi-

TABLE 6.3. Nutritional Comparison of the Costs of Imported and Locally Produced Foods, 1989

Item	Kilocalories/ EC$1.00	Grams of protein/ EC$1.00
Imported		
Rice	1,734	32.0
Flour	2,751	79.3
Locally produced		
Dasheen	782	13.4
Plantain	688	6.3

Sources: Field survey; Caribbean Food and Nutrition Institute 1974; and Grossman 1993.

cal conditions of production and marketing in St. Vincent and in the countries that are the providers of its food imports. The United States is St. Vincent's major supplier of wheat and chicken, the two main sources of imported carbohydrates and animal protein, respectively (SVG Statistical Unit files). Agribusiness in the United States benefits from economies of scale, mechanized production, substantial university and extension research and support, and favorable U.S. government subsidies and export policies, enabling corporations to sell products overseas at low prices. In contrast, Vincentian farmers who supply the Kingstown market receive minimal government assistance and employ primarily human labor and comparatively low levels of technology on small, scattered, steeply sloped plots. Mirroring the situation in much of the English-speaking Caribbean, the Vincentian pattern, a reflection of the historical pattern of property relations and the traditional lack of government interest in domestic food production, results in comparatively high production and marketing costs (Long 1982; McIntosh and Manchew 1985; Thomas 1988). In addition, the island's only mill that processes flour and rice benefits from a variety of tax incentives provided by the Vincentian government for industries locating in St. Vincent and the Grenadines, including the duty-free importation of these two key components of the local diet. Also, government price controls affect retail prices of imported foods but not of most local food crops, thus imparting more stability to the prices of imported items.[16]

Other reasons exist for the popularity of imported foods. One is taste: children, in particular, prefer imported foods over local ground provisions. Mothers from both rural and urban areas have informed me that children complain that consumption of provisions produces a "heavy," unpleasant feeling in their stomachs; as one women noted in referring to provisions, "children make 'noise' if they get 'hard food' [provisions]." Moreover, imported foods are more

FIGURE 6.4. Percentage Increase in Cost of Foods, St. Vincent, 1964–1990

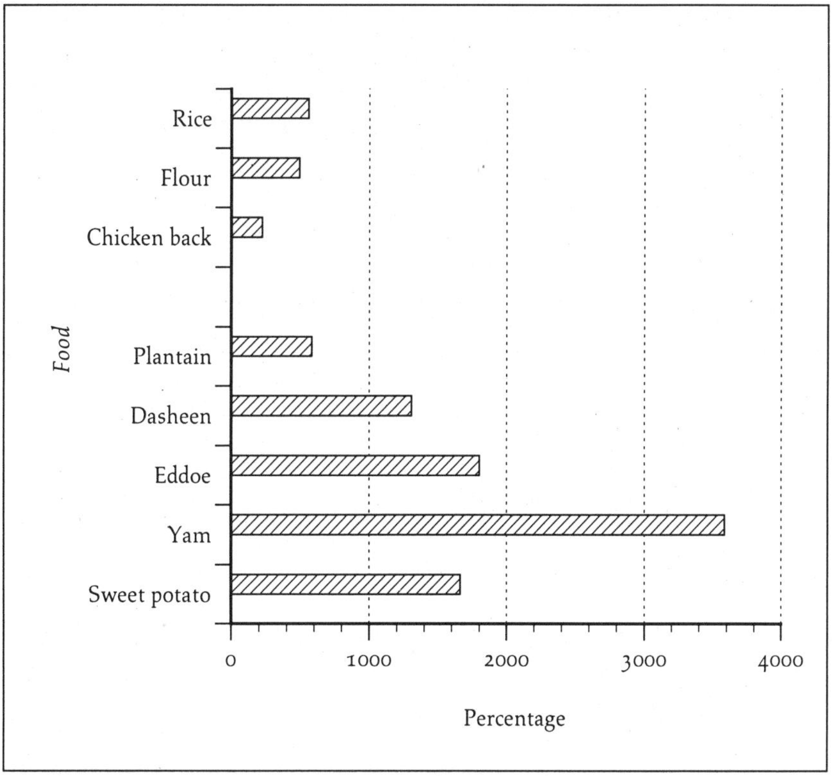

Source: SVG Statistical Unit files.

convenient and quicker to prepare. They also have a longer shelf life. Many complain that the shelf life of ground provisions has declined over the last ten to twenty years, making them less appealing to purchase. One likely reason for the decreased shelf life is the dramatic increase in the use of synthetic fertilizers on food crops during this period (Browne 1985); such agrochemicals increase both the size and water content of provisions, making them spoil more quickly than before. Interestingly, the considerations of taste, convenience, and shelf life that figure so prominently today have actually been long-standing influences on food import dependence, as earlier works indicate (Engledow 1945; Jolly 1947).

Thus, even though a sizable domestic market exists—from both urban and rural areas—that market is supplied primarily by imported foods. People in urban areas tend to consume local food crops purchased from the Kingstown market more on weekends, when meals are traditionally larger and have more varied dishes. Thus, incentives to expand food production for the Kingstown market in banana-growing areas were limited in the 1980s and early 1990s. At that time, it was not uncommon to find market vendors returning to their villages with unsold provisions.[17]

The changes in Restin Hill in 1994 are illustrative of the significance of market conditions for food crops. Drought in St. Vincent in 1994 led to declines in food crop yields, causing prices in the market to rise. At the same time, farmers in Restin Hill were becoming more and more frustrated with falling banana prices and the increased labor costs associated with the cluster pack system, resulting in a decrease in banana acreage in the village of roughly 20 percent. Given the altitudinal range exploited, Restin Hill farmers were less seriously affected by drought than most others in St. Vincent, enabling them to increase the areas planted in food crops to take advantage of the higher market prices. Thus, the lure of higher prices, coupled with the downturn in the banana industry, finally provided the incentive to expand food production for the Kingstown market.

Starting in 1995, another change was noticeable. Considerably more women from throughout St. Vincent were marketing food crops in Kingstown. The government official in charge of the market indicated that as many as six to seven hundred women were now selling food crops on Saturdays, compared with a maximum of five hundred in the late 1980s (Edward Ross, pers. comm., 18 November 1995). While the continuing decline of the banana industry contributed to increased marketing of food crops, another reason for the expansion in the number of sellers in the Kingstown market was growing weakness in the export market for local food crops.

Villagers also sell their commodities to traffickers, who are mainly women and who export the items to nearby islands. Trinidad to the south is the primary destination for these traders, who have been shipping Vincentian commodities there since the 1800s. Today, provisions are the main commodities shipped to Trinidad, though significant amounts of plantains and citrus fruits are also part of the trade. Barbados to the east is the second most important market for such crops.

Growth of the Trinidad market has been closely linked to the pattern of provision production on St. Vincent, because most of Vincentians' provisions are, in fact, exported there. For example, the SVG Department of Agriculture (1984: 9) reported that, except for dasheen, 80 percent of the root crops produced on St. Vincent were exported in 1980. Demand for St. Vincent provisions grew considerably in the late 1970s, fostered by the boom in Trinidad's petroleum industry. Indeed, much of the growth in provision production on St. Vincent from the late 1970s to the mid-1980s was destined for that market (Figure 6.5).[18] The increased production of provisions depicted in Figure 6.5

FIGURE 6.5. Provisions Production, St. Vincent, 1977–1989

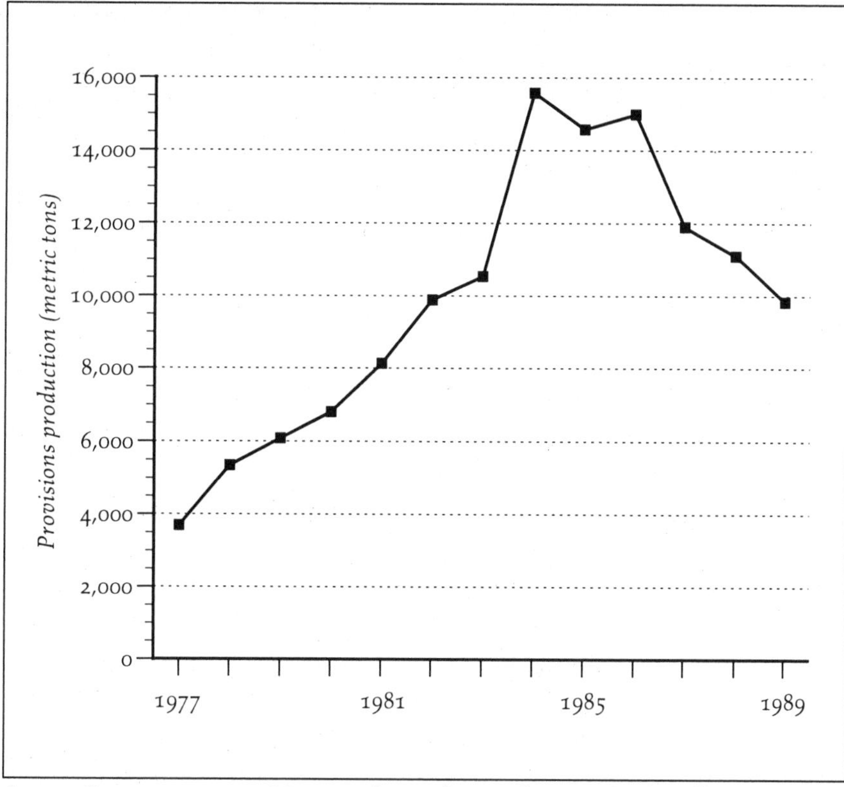

Sources: Grossman 1993; SVG Ministry of Agriculture, Industry, and Labour files.

came from areas not suitable for banana production and with limited alternative markets for their crops. Peasants in Restin Hill and other banana-producing areas also sold provisions to traffickers but did not expand their cultivation of provisions in the 1980s because of numerous problems in the trafficker market.

In trafficking, no guaranteed market exists, and sometimes villagers cannot find buyers for all the crops they want to sell—problems typical in open-market relations. When traffickers need produce, they contact farmers in the countryside, and growers, in turn, seek out traffickers when wanting to sell. After an agreement is reached on prices and the amounts that the traffickers will buy, farmers eagerly await instructions concerning when to harvest, as flexibility in timing is possible because the crops can remain in the ground for some time.

Marketing through traffickers entails considerable risk and uncertainty. Price fluctuations, similar to those in the Kingstown market, discourage farmers. The possibility of not being paid is an even greater risk. Many traffickers obtain crops from peasants on a "trust" or credit basis, promising to pay for them when they return from selling their crops overseas. But some fail to keep their promises, claiming bankruptcy because of unprofitable marketing transactions in Trinidad. Peasants also complain of sometimes having to wait two or three months before receiving their payments—situations in marked contrast to the regularity of payments from the SVBGA for bananas.

Despite these disadvantages, peasants in banana-growing areas continued to sell to traffickers instead of relying exclusively on bananas. Diversification of production reduces the risks associated with reliance on a single crop. Also, depending on their relative location and environmental characteristics, not all plots in banana-producing regions are, in fact, suitable for bananas. Also, those households constrained by labor shortages tend to rely on less labor-intensive crops, such as provisions and plantains, which are often sold to traffickers. Furthermore, given the small acreages that many peasants control, they attempt to maximize output from their limited holdings by intercropping bananas and food crops.

The trafficker market, however, started to decline in the latter part of the 1980s. In response to falling oil revenues and a variety of other economic problems, Trinidad and Tobago devalued its currency in 1985, 1988, and again in 1993, making the exporting of provisions there from St. Vincent much less profitable than before. The marked decrease in provision production starting in 1986 (Figure 6.5) was in response to the decline in the Trinidad market, not to competition from banana cultivation.

Exports of provisions continued to fall in the first half of the 1990s (Figure 6.6).[19] Not only did Trinidad's economic problems dampen demand for Vincentian exports, but the government there also encouraged more efforts at provision production at home, further limiting the trafficker market ("Women Traders of the Caribbean" 1990). In fact, the temporary upswing in Vincentian provision exports in 1992 reflected a decline in provision production in Trinidad that year. Problems intensified for growers focusing on exports, as a pink mealybug infestation in Trinidad in 1995 led initially to a short-term ban on the importation of Vincentian provisions and then to the imposition of a new regulation requiring that all provisions imported into Trinidad be washed prior to shipment, adding a considerable labor burden to Vincentian farmers. And in 1996, to make matters worse, Barbados terminated food imports from St. Vincent in response to reports that the pink mealybug had also infested Vincentian gardens. Thus, the growth in the number of women selling in the Kingstown market in the mid-1990s reflected not only weaknesses in the banana industry but also the decline in the trafficker market, as fewer provisions were leaving the country and growers shifted their focus to the home market.

This analysis of market problems for local food crops is indispensable in

FIGURE 6.6. Provisions Exports from St. Vincent, 1989–1994

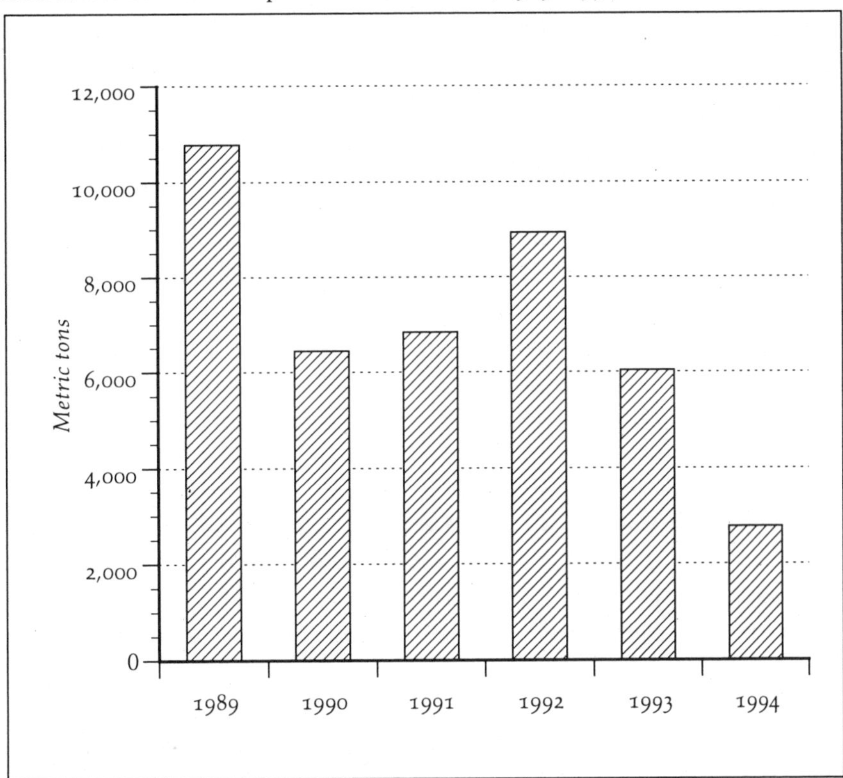

Source: SVG Statistical Unit files.

evaluating the relationship of banana production to increased food imports. Clearly, food insufficiency related to banana production is not the reason—at either the local or the national level—that food imports are growing in St. Vincent. At the local level in Restin Hill, people rely primarily on imported foods not because food crops are unavailable but because of dietary preference, convenience, and cost. Food crops are available, but people prefer to sell the majority of their provisions, plantains, and greens. At the national level, production of provisions boomed from 1977 to 1986 in areas not producing bananas at the same time that food imports were growing. Provisions were available to supply a large part of the domestic market, but Vincentians favored imported flour and rice.

To complicate the issue of the impact of banana export production, bananas have become an important addition to the food supply at both the local and national levels (see also Trouillot 1988). Vincentians consume bananas unsuitable for export at their major meals and as snacks, in green and ripe forms, respectively. Bananas are also sold widely in the Kingstown market, where the ripe state is particularly popular. Although official figures on the amount of bananas consumed in St. Vincent do not exist, I estimate conservatively that Vincentians consumed roughly 1.1 million pounds of bananas (net weight) per year in 1988 and 1989.[20]

Interestingly, bananas have entered traditional food markets in another manner. With the decline in banana export prices, levels of care in banana fields started to drop in 1993, leading to an increase in the production of substandard bananas unsuitable for export. Moreover, farmers have become more selective in determining which fruit to pack with the cluster pack method, leaving more rejects in their fields. This combination created a larger supply of available bananas that were not exported to the United Kingdom, and traffickers quickly took advantage of the situation by increasing exports to nearby islands, especially Trinidad, Grenada, and Barbados in 1994.[21] Consequently, exports of bananas to the Caribbean region rose from an annual average of 1,119 metric tons between 1988 and 1991 to 4,486 tons in 1994 (SVG Statistical Unit files). Thus, although the majority of bananas were exported to the United Kingdom, many also entered the realms of subsistence production, the Kingstown market, and the regional economy—complementing local food production focused on these arenas.

Bananas and Local Food Production

The agricultural landscape of Restin Hill has changed dramatically since World War II. A visit to the village in the 1950s and 1960s would have revealed fewer bananas planted for the export market, larger areas cultivated in ground provisions and plantains, more cattle supplying milk for households and manure for fields, and more land under fallow. What has led to the decline of both local food crop production and cattle raising? The most obvious answer would appear to be the expansion of banana production during the same period, but the issue is more complicated. Here we need to follow the focus of political ecology and examine the political-economic and environmental contexts of agricultural production to understand the role of banana export contract farming in changing patterns of food production.

When the growth of export agriculture is accompanied by declining local food production and rising food imports, the general assumption in the development literature is that export agriculture has undermined local food production, leading to insufficient food supplies and thereby requiring increased food imports. Researchers emphasize specifically that such negative impacts on food production result from conflicts in the productive sphere (Wisner 1977; Porter 1979; Grossman 1984a; Bassett 1988b; see also Nietschmann 1973). Contract farming is particularly relevant to such conflicts. Capital and the state may attempt to intervene in the peasant production process by dictating the technologies to be used, the amounts and timing of labor inputs, and the areas devoted to contracted export crops (Bassett 1988b). Thus, the key issues here are whether control by such forces is responsible for regimenting production patterns and whether such regimentation leads to conflicts in the productive sphere that are the primary cause of the decline in local food production.

The first issue to consider is whether significant seasonal conflicts exist. That is, are the labor demands of banana production so markedly seasonal that they detract from the time available for food production at certain critical times of the year? In some contract-farming schemes, capital and the state are able to control farmers' activities rigidly, ensuring that the cultivation and harvesting of the contracted commodities occur during those seasons that are optimal for the crop, thereby forcing farmers to allocate labor to their food crops during less favorable times of the year, which lowers food crop productivity (Bassett 1988b; Little and Watts 1994). As we have seen, however, such rigid control in the phases of cultivation is absent in the case of banana production in St. Vincent, in part because of the low ratio of extension agents to farmers.

Moreover, an environmental reality—the nature of the banana plant itself—makes strict supervision difficult. Banana plants do not all grow at the same rate, and divergences in growth and production among plants in a garden increase as the field becomes older. Furthermore, growth patterns of bananas within St. Vincent also vary spatially in response to rainfall conditions, which are not only influenced by elevation but also highly localized. Thus, it is not feasible to enforce a rigid, uniform scheduling of activities that would be applicable to all banana farmers on the island.

Some seasonality does exist in banana production. Price incentives and seasonal rainfall patterns encourage Restin Hill farmers to plant somewhat more bananas from May to October and harvest somewhat fewer fruit from January to May than at other times of the year. The banana plant is mostly water and thus thrives when rainfall is regular and adequate but slows its growth and rate of production when under moisture stress, as is typical during the dry season. However, the altitudinal range exploited by the villagers helps dampen such impacts of seasonality.

People harvest bananas throughout the year either weekly or every other week. Harvesting activity is a valid indicator of seasonality in the system because it is the most time-consuming aspect of banana production. A comparison of monthly banana sales (which reflect patterns of harvesting) of the sample

FIGURE 6.7. Monthly Banana Sales, Sample of Households, 1989

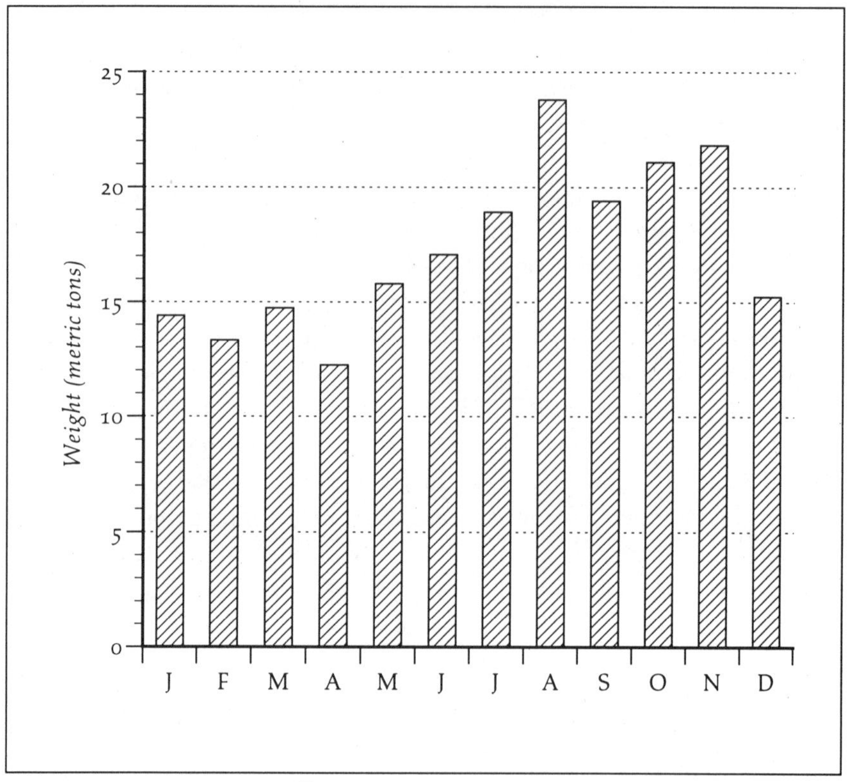

Sources: Grossman 1993; SVBGA files.
Note: N = 15 households

members with SVBGA banana exports (which reflect harvesting nationwide) during 1989 reveals that seasonal patterns at both the village and national levels were not extremely marked (Figures 6.7 and 6.8).[22] The somewhat greater degree of seasonality in Restin Hill banana sales compared with the national pattern likely results in part from the effects of the tropical storm that hit the island in September 1987. Damage from the storm was concentrated in the Marriaqua Valley and forced many farmers there to replant their fields prematurely. Such widespread replanting in the valley at the same time produced a more seasonal pattern of harvests than is normal.

Similarly, the time spent on all banana-related activities does not exhibit marked seasonality (Table 6.4), reflecting not only the pattern of harvesting but also that of planting. Peasants normally do not plant an entire field of bananas at once but instead plant different sections at different times to help distribute the workload more evenly throughout the year.

FIGURE 6.8. Monthly Banana Exports by SVBGA, 1989

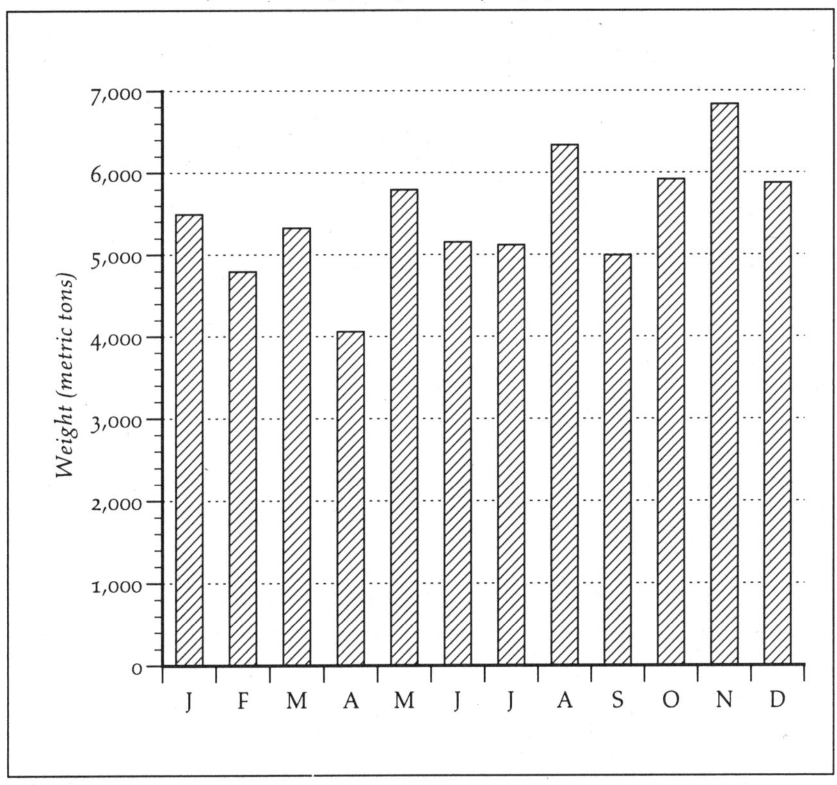

Sources: Grossman 1993; SVBGA files.

The pattern of time spent on food production is also not very seasonal. Peasants can plant provisions throughout the year because of the more adequate rainfall at higher elevations in the dry season. Nevertheless, overall plantings of provisions do tend to decline somewhat during the drier months. Yet the pattern of "greens" cultivation complements that in provision production, as most of the effort in cultivating cabbage and tomatoes, two very labor-intensive crops, occurs from January to June. In any case, most people in Restin Hill intercrop bananas and food crops when replanting their fields—a pattern found elsewhere in St. Vincent and the rest of the Windwards.[23] Thus, seasonal conflicts between banana production and food crop cultivation are not the cause of declines in local food crop production.

The related issue is whether the total overall time required for banana production is so high that it substantially reduces the time available for food crop cultivation throughout the year. Banana production is certainly very labor intensive; it required approximately 150 people days per acre per year in on-

TABLE 6.4. Seasonality of Banana-Related Activities, Sample Households

Month	Percentage of Time Spent on Banana-Related Activities
1988	
November	7
December	8
1989	
January	7
February	4
March	8
April	5
May	6
June	6
July	6
August	11

Sources: Time allocation study of twenty-four sample households; Grossman 1993.

farm and off-farm activities in the late 1980s. Production in fields with local food crops is usually less labor intensive, although those having large amounts of greens require considerable time.

The issue of time availability is crucial, because many farmers complain about labor shortages. But the conclusion that time constraints in agriculture are caused primarily by the labor-intensive nature of banana production overlooks more complex issues. A political-ecological perspective helps focus on the critical issue of labor availability in relation to differential control over resources.

Although it might be expected that households with limited amounts of land—two acres or less—would not have labor problems in agriculture because of the small areas involved, such is not the case. As Collins (1987) and Zimmerer (1993) note in their studies in South America, peasants with small landholdings may experience time constraints in agriculture because their small holdings do not produce enough to support their households, thus necessitating their participation in wage labor, which in turn reduces the time available for cultivation. Such an argument, focusing on the distribution of resources as a fundamental cause of labor shortages in agriculture, is certainly relevant to the present case. In Restin Hill, those with smaller holdings are clearly more likely to be engaged in wage labor (Table 4.2), which reduces their time available for agricultural pursuits. In the sample households with holdings of two acres or less, wage labor was especially time consuming for men; adult males (those fifteen years and older) spent 33 percent of their time working for wages, with male heads of such households spending the most time—49 percent—on this activity. Banana production is not the root of the problem: lack of adequate access to land is the key problem for many farmers, leading to time constraints in agriculture.

Farmers with small holdings attempt to manage their time constraints in a variety of ways. When labor demands are high for a day or two, as they are during banana harvesting or soil tillage, people can obtain extra labor through reciprocal labor exchanges—swap labor and help relations. Recourse can also be made to hiring labor, though its use is much less frequent than among those households with larger holdings. Another response is simply to delay activities; for example, banana fields past their prime that should have been cut down and replanted are left standing for several more months.

For households controlling more than two acres, time constraints can also be a problem. To help alleviate such shortages, these households hire labor more frequently than those with more limited holdings. Obtaining adequate wage labor assistance, however, can be difficult because of the overall shortage of agricultural wage laborers.

Such shortages of available wage labor in agriculture have always been a problem but intensified in the 1980s. More teenagers are now enrolled in secondary school than previously, which reduces the available labor pool. In addition, as the Vincentian economy grew in the 1980s, more young men were able to obtain off-farm employment in a variety of trades. Banana production is also partly responsible for this shortage. Farmers recall that before the rapid expansion of banana production in the 1980s, hiring wage laborers for assistance in agriculture was easier. As banana production became more rewarding financially in the 1980s, young men and women, who were a traditional source of agricultural wage laborers, were drawn back into farming for themselves, seeking to obtain small parcels of land (primarily through gift, temporary-use rights, rental, and sharecropping) to grow bananas on their own. Had they not obtained land for banana cultivation, they would likely have supplied more wage labor in agriculture. But such a trend has not been negative for food production because these young farmers also tended to intercrop bananas with local food crops to maximize output from their small plots.

In the context of limited wage labor availability, the comparative willingness of wage laborers to work in banana cultivation as opposed to food production is a crucial issue. The task for which farmers hire labor most frequently in local food production is soil tillage. The labor pool for this task is limited primarily to young men, but many of them are hesitant to work for wages tilling the soil, because the work is very hard and has low status. In contrast, hiring labor in banana production is easier. The task for which banana growers hire labor most frequently, carrying banana boxes, is much less demanding physically than soil preparation. In addition, the potential labor pool for carrying boxes is much larger, as both males and females, adults as well as children as young as ten years old, are hired for the task. Consequently, shortages of agricultural wage labor inhibit food production more than banana production.

Another dimension that lessens potential time conflicts should be noted—the weekly scheduling of events. The most labor-demanding aspect of banana production—harvesting—traditionally occurred on Tuesdays and Wednesdays, though the SVBGA shifted the time to Mondays and Tuesdays beginning in 1994. In contrast, food harvesting is particularly intense on Thursdays and Fridays, when people prepare for marketing the next day.

The second set of issues in examining the significance of conflicts in the productive sphere relates to spatial patterns of land use (Map 4). The relative location of agricultural activities thus requires investigation. The frequently made generalization that the expansion of export cropping pushes local food crops farther from villages into more marginal environments (see Grossman 1984a; Barbier 1989) needs modification in banana contract farming. Models indicating that export cropping is located closest to villages make two critical assumptions that can be questioned in the present case: first, that distance from house to garden is the key variable influencing farmers' spatial decisions and, second, that export cropping and local food cropping are distinct spatially.

Regarding the first assumption, peasants are always concerned about distances from their house to gardens, but banana growers are even more concerned about distances from their gardens to the nearest roads. Whereas walking distances between houses and banana gardens of the sample are up to forty-five minutes, those from banana gardens to roads are eighteen minutes or less. Eighty-nine percent of the area under bananas lies within a nine-minute walk of a motorable road. This spatial pattern reflects the need to carry bananas weekly or biweekly on harvest days from fields to roads (where the boxes are loaded onto trucks), the high bulk produced per unit area annually, and the need to export high-quality fruit. Bananas are easily bruised, and bruising reduces fruit quality. In this context, the environment is a crucial variable; the rough topography and slippery paths over which farmers must carry their boxes of harvested bananas increase the likelihood of bruising, thus placing a greater premium on short distances between field and road than between field and house.

The second assumption can also be questioned because bananas and food crops are often grown together in the same gardens; on the majority of land with food crops (69 percent), these crops are either intercropped with bananas or are planted alone in one section of a field that has bananas on it elsewhere. The distribution of distances between farmers' houses and their banana and

MAP 4. Food and Banana Gardens, Sample of Households

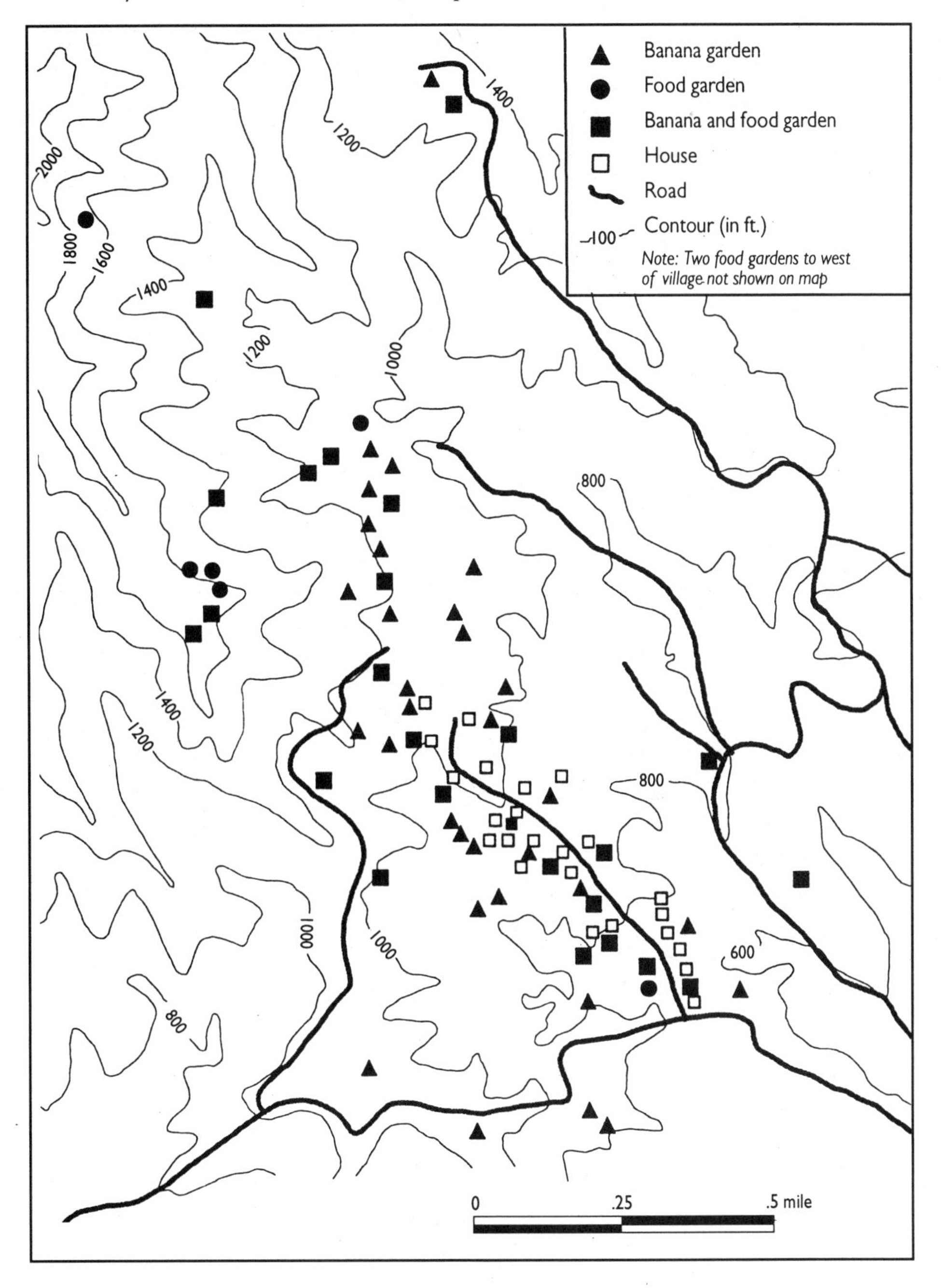

TABLE 6.5. Relationship between Land Use and Distance from Residences, Sample Households

Time Distance from Residences to Gardens (minutes)	Percentage of Land under Bananas[a]	Percentage of Land under Local Food Crops[a]
1–9	34	24
10–19	41	28
20–29	7	14
30–39	0	5
40–49	17	18
50–59	0	11

Sources: Survey of gardens of sample households; Grossman 1993.
[a]Includes land intercropped with both bananas and local food crops

food plots does not indicate marked spatial differences, though bananas are, on average, somewhat closer to the village (Table 6.5). Such differences are not significant enough to account for the decline in food production.

Generalizations that export cropping also pushes food production onto more marginal lands while occupying the best lands also require reexamination, in part because of the pattern of food crops and bananas usually being planted in the same fields. Also, as the area under bananas expanded, particularly in the 1980s, it was not the local food crop gardens exclusively that were pushed onto more marginal lands. Rather, farmers planted new banana fields (often intercropped with local food crops) on steep, marginal lands at the tops of hillsides previously uncultivated or in very long fallow and highly susceptible to erosion. The proximity of such sites to roads was a more important factor influencing location than were concerns about steep slopes. Although banana gardens are on land that is, on average, less steep than that in food gardens, much of the land under bananas is also quite steep (Table 6.6). Part of the differences in the distribution of slopes reflects spatial patterns within plots that have both bananas and food crops but in which the crops are planted on different sections of the same hillside; villagers tend to plant food crops at the top portion of hillsides and bananas on the lower portion for environmental reasons: to limit exposure of bananas to winds, which can topple the top-heavy plants, and to prevent both fertilizers and pesticides applied to bananas from being carried by runoff downhill onto food crops.[24]

The other spatial issue is the extent to which banana production competes directly with local food crops for the limited land available. The majority of Restin Hill villagers interplant food crops with bananas when replanting a field but harvest most of the food crops (except tannia) within nine months after

TABLE 6.6. Relationship between Land Use and Slope, Sample Households

Slope (degrees)	Percentage of Land under Bananas[a]	Percentage of Land under Local Food Crops[a]
1–9	19	10
10–19	33	17
20–29	37	33
30–39	11	35
40–49	0	5

Sources: Survey of gardens of sample households; Grossman 1993.
[a]Includes land intercropped with both bananas and local food crops

planting. Before they replant, farmers usually allow bananas to remain in the field for another one to two years, producing two or three ratoon crops, during which time local food crops cannot be interplanted because the tall bananas create too much shade. Growers nevertheless manage to produce enough local food crops to supply their subsistence and marketing needs by planting different portions of a field at different times of the year so that food crops interplanted with bananas will be available for much of the year. In some cases, people continue to replant a field or section of a field only in bananas, without intercropping for a period of many years. But such households also often reserve a portion of that plot or another field where they plant no bananas so as to ensure a regular supply of local food crops.

One clear case of conflict does exist, however—that between banana production and cattle raising, which has caused declines in both the "pen patch" and "tie-out" systems. The pen patch system has suffered from a decrease in available feed to cut; banana farmers now use the herbicide Gramoxone (paraquat), which desiccates garden weeds that used to be fed to cattle, and the tall bananas tend to shade out grass strips traditionally planted as fodder along garden boundaries. The tie-out system has suffered as the area of fallowed land has been reduced by the expansion of banana cultivation. Moreover, the traditional pressure to raise cattle as sources of much-needed manure for fields and milk for diets has declined drastically. Villagers now rely mostly on synthetic fertilizers obtained primarily from the SVBGA and powdered milk purchased from stores to meet these needs.

To conclude this analysis of the significance of possible conflicts in the productive sphere between local food production and banana contract farming, another complicating factor must be considered. Certain aspects of banana production actually *encourage* local food crop cultivation. Villagers assert that when bananas are intercropped with food crops, bananas usually ripen one month earlier and bear larger bunches than when planted in monocrop stands. The benefits from interplanting reflect several features of food crop cultivation: farmers weed food crop gardens more thoroughly than bananas planted alone in monocultural stands, an important consideration because banana yields are particularly affected by the thoroughness of weeding during the early stages of growth; villagers till the soil completely when planting food crops, which improves soil aeration and percolation and facilitates root penetration for the banana plants; and leaf cover from the dasheen crop helps limit weed growth in the field. Also, on the steep hillsides on which people must grow much of their crops, peasants often prefer to plant bananas with food crops to limit the loss of fertilizers and pesticides from runoff. When people plant bananas alone, they employ minimum tillage—disturbing the soil surface only to dig holes for the widely spaced banana plants while leaving the rest of the field uncultivated. Such flat surfaces on steep gardens would allow rainfall to wash agricultural chemicals rapidly off the plots. In contrast, people till the soil completely when planting food crops, constructing ridges and furrows along contours in the "range and fork" and "range and cover" systems. This pattern of tillage helps prevent the heavy doses of fertilizers and pesticides applied to bananas from washing off their fields, because the agrochemicals are trapped in the furrows during rainfall; indeed, this preference for intercropping to limit the loss of agrochemicals helps to account, in part, for the high percentage of land under food crops on steep slopes (see Table 6.6). In addition, when bananas are intercropped with food crops, the latter benefit from the synthetic fertilizers obtained for bananas. Furthermore, the light Vincentian soils do not provide as secure an anchoring for bananas as do heavier clay soils found on some other islands elsewhere in the region; consequently, the corm at the base of the banana plant tends to move upward in the soil with each successive ratoon, making the plants less stable as they get older. As a result, farmers tend to replant their fields after only a few ratoons, which facilitates frequent intercropping with food crops. Finally, people plant food crops as insurance against possible loss of banana production owing to periodic windstorms and hurricanes, always potential threats in the Windwards.

Ultimately, a focus on conflicts in the productive sphere implies that food crops were the only alternative to bananas. In some areas, such as in Restin Hill, that assumption is at least partially correct, as food crop cultivation was more extensive before the expansion of banana production, although villagers also grew arrowroot and, before that, sugar cane. But not all areas of St. Vincent were as diversified in production as was Restin Hill. In many places, considerable dependence on classic export crops was evident among peasant growers. For example, the Agricultural Department's annual review for 1938 (SVG Agricultural Department 1940: 28) comments on "the necessity which forces the average peasant to maintain practically the whole of his small area of land under either arrowroot or cotton in order that he may obtain the maximum cash return to meet his relatively heavy financial commitments." Although not all Vincentian peasants focused on these exports crops to the extent indicated in the report, many did. Consequently, in such areas, bananas mainly replaced other classic export crops—cotton, arrowroot, and sugar cane. The Department of Agriculture's (SVG Department of Agriculture 1959: 15) report in 1957 supports such a view: "The decrease [in sugar cane] was due largely to the extended cultivations of bananas especially by peasant proprietors. . . . Peasants continued to abandon other crops especially arrowroot which were of lower economic value to plant bananas." Indeed, the replacement of one export crop by another in St. Vincent has occurred repeatedly over time.[25] For example, arrowroot expanded at the expense of cotton on peasant holdings (Spinelli 1973).

Moreover, historically in St. Vincent, competition within peasant holdings between local food crops and traditional export crops—sugar cane, cotton, arrowroot, and most recently bananas—has usually occurred in the context of the relative strength of their *export* markets. Agricultural Department reports in the 1930s and 1940s described competition between sweet potatoes and cotton for the same land, as both thrived in drier parts of the island—competition that was based on the strength of their respective export markets, with the relative success of sweet potatoes at the time dependent on demand from its market in nearby Trinidad (SVG Agricultural Department 1945: 2). Similarly, today, the viability of local food production in relation to banana contract farming is linked, to a considerable degree, to the nature and health of the Trinidad market and to the comparative returns from the banana export market, not to conflicts in the productive sphere generated by the control exercised by capital and the state over the peasant labor process.

The Political-Economic Contexts of Agricultural Production and Marketing

The contrasting political-economic contexts of producing and marketing local food crops and export bananas have been essential forces influencing changing patterns of land use. In particular, marketing policies and intervention by capital and the state in the peasant labor process have given banana production substantial advantages over local food crop cultivation, thereby contributing to the decline in local food output.

Historically, governments in the English-speaking Caribbean have given much less support to the marketing of food crops than to export crops (Long 1982; Axline 1986; Thomas 1988), and the current situation in St. Vincent and elsewhere in the Windward Islands is no exception. With minimal assistance from the government in marketing their local food crops, farmers are plagued with substantial price fluctuations and limited and uncertain demand, problems inherent in open markets. In contrast, banana growers have had a guaranteed market provided by the SVBGA, with a regular weekly or biweekly income. They have also benefited from a much greater degree of stability in relation to pricing. Each year from 1987 to the end of 1992, the SVBGA had set, in advance, fixed prices to be paid for bananas for the coming twelve-month period. At other times, even when prices have dropped markedly in the United Kingdom, the SVBGA has often muted the impacts of such price swings by subsidizing the price to growers, an action made possible by either drawing on its own reserves or borrowing from banks. The more attractive market provided by the SVBGA must be understood in light of the producers involved. Whereas peasants grow both bananas and provisions, more politically influential, larger-scale capitalist farmers tend to focus on banana production. This merging of interests of small- and larger-scale growers in the SVBGA membership is politically significant; policies supportive of the banana industry serve the interests of both groups of producers at the same time. Moreover, the market policies of the British government also helped support the price of bananas, with the result that for much of the 1980s and early 1990s, banana prices were comparatively high, though the creation of the Single European Market in 1993 has eroded such benefits.

The Vincentian and British governments have had their own particular interests in supporting the expansion of export banana production. British encouragement of the industry since the 1950s can be interpreted as a form of foreign aid designed to bolster the economies of the islands while also serving British interests. The Vincentian government has provided support because of welfare considerations, the political significance of the large size and composition of the SVBGA membership, the overall stimulative effect of banana production on the Vincentian economy and hence on government tax revenues, and the essential need to earn foreign exchange in light of perennial, growing trade deficits.

Intervention by capital and the state in the peasant labor process—a hallmark of contract farming—has also helped make bananas preferable to local food crops. Where such intervention does negatively affect food production, it is normally associated with its contribution to conflicts in the productive sphere. But in this case, a more novel relationship is involved. The continuing pressures to improve fruit quality from Geest, the British government, the British market, and, more recently, the EU have led to mandated technological innovations in harvesting and packing; these have provided banana production another advantage over local food crop production that is related to the perennial problem of crop theft. One of the traditional reasons for the limited areal extent of cultivation in the Eastern Caribbean generally was the high incidence of "praedial larceny" (Shephard 1945; Rubenstein 1975, 1987). Engledow (1945: 12) described the seriousness of this long-standing problem in his supplement to the Moyne Commission Report: "No circumstance affecting home-grown food production is more widely or more forcibly brought to notice than what in these Colonies is known as praedial larceny. It is the theft of crops or even of small or young animals. . . . It affects agriculture in many ways. . . . Praedial larceny is, indeed, a profound handicap to agricultural development." Similarly, a report on agriculture in St. Vincent (SVG Agricultural Department 1953: 17) lamented "the wave of praedial larceny which prevails and so discourages peasants from growing more food." A thief could easily go into someone else's food garden under cover of night, dig up provisions, put them in a sack, and carry them to Kingstown the next day to sell to hucksters without being discovered. In contrast, technological innovations in harvesting and packing bananas have made them less susceptible to theft than food crops. St. Vincent and the other Windwards have been forced to adopt a succession of complicated, intricate, labor-intensive processing methods—field packing, paco pack, and, most recently, cluster packing. Given the increasingly intricate, detailed steps required in these systems, stealing bananas at night, processing and packing them according to specifications, and then selling them to the SVBGA the next day are extremely difficult—yet one more advantage of banana production over local food crop production.

Seven

The Environmental Question

Much of the literature on political ecology focuses on the political-economic influences on environmental degradation. Thus, work informed by this perspective emphasizes how capital and the state affect human-environment relations, especially by impoverishing small-scale producers (Schmink and Wood 1987; Bassett 1988a; Durham 1995; Painter 1995). Such analyses of structural forces have certainly helped to elucidate processes of change in resource-use patterns.

But a complementary view has also emerged within political ecology that focuses on the significance of discourse, perceptions, beliefs, and associated behaviors (see Zimmerer 1991, 1996a; Blaikie 1995a,b; Jarosz 1996b; Peet and Watts 1996). While such dimensions are not portrayed in the literature as being determined in any simplistic manner by structural forces, the categories chosen for analysis still reflect differences in economic power and wealth. Thus, views of government officials concerning such issues as deforestation or soil erosion are contrasted with those of peasants, or perspectives of older villagers are compared with those of younger ones, or the role of gender is highlighted. Clearly, such differences are fundamental influences on perceptions, beliefs, and behavior and are relevant to understanding human-environment relations. And it is logical for political ecology to emphasize such contrasts, given its focus on political economy. But I suggest that we need to extend such analyses by appreciating the significance of the environmental dimension itself—in particular, the environmental rootedness of agriculture. Specifically, the environmental context of farming helps generate substantive interpersonal variations in perceptions, beliefs, and behavior among individuals *within* the same groups and classes, not just between such groups. And such intragroup variations have fundamental implications for understanding pesticide misuse.

Surprisingly, few studies of contract farming focus on environmental impacts (Little 1994: 246), though exceptions exist (see Murray and Hoppin 1992; Murray 1994; Morvaridi 1995). But the topic is particularly appropriate because capital and the state attempt to regulate production processes in contract farming, including intensive agrochemical use. In particular, pesticide use, because of its link to the production of high-quality produce, is usually a crucial dimension in contract-farming enterprises.

But just as capital and the state have difficulty achieving thorough control over the labor process generally in banana production, so too is control over pesticide-use practices problematic—an issue directly relevant to the appropriateness of the concept of deskilling. Although capital and the state attempt to instill particular, uniform regimes of agrochemical use, what actually happens at the local level is much more complex. Similarly, the conception of standardized technologies in contract farming characteristic in the literature on the new international division of labor in agriculture (see Sanderson 1985b, 1986b; Bonanno et al. 1994) glosses over the considerable variability in the implementation of such technologies. In this chapter I focus on the issue of pesticide use and misuse, exploring the extent and significance of variability in production patterns.

Certainly, pesticide misuse is not the only environmental problem associated with banana production. Soil erosion is another issue (Thomson 1987), but pesticide problems are more closely associated with the banana industry. Indeed, given the rugged terrain on the island, the threat of soil erosion has been an ever-present danger for all cropping systems; in fact, soil loss due to agriculture has been widespread in the Caribbean generally since the time of slavery (Watts 1987). The problem of soil erosion has been exacerbated by impoverishment associated with a long history of unequal control over land, which has evolved into the current pattern of peasant cultivation on steep hillsides. In this context, the banana industry has been but one in a succession of commodity systems assaulting the soil; indeed, soil erosion under cotton cultivation previously was more severe than it is in banana growing in the present era. In contrast, although pesticides were used before the start of the contemporary banana industry in the Windwards, the extensive use of these agrochemicals today is clearly linked to the industry's growth and ascendancy.

Understanding Pesticide Use

A prominent feature of contemporary agriculture in St. Vincent and elsewhere in the Caribbean, whether or not contract farming is involved, is increasing dependence on synthetic pesticides (Gooding 1980; Pollard 1981; Collymore 1984; Caribbean Agricultural Research and Development Institute 1985; Murray and Hoppin 1992; Murray 1994; Patterson 1996), a trend paralleling that in many other developing countries (Weir and Schapiro 1981; Bull 1982; Goldberg

1985; Thaman 1985; Tait and Napompeth 1987; Thrupp 1988, 1990a,b, 1994; Wright 1990). Understanding patterns of use and misuse of agrochemicals is essential because they can have significant short- and long-term deleterious impacts on human health and the environment. Problems of pesticide misuse are more severe in developing countries than in advanced capitalist countries, and it is well known that the incidence of pesticide poisoning in the former is much higher (Bull 1982).

A prominent theme in the literature is the role played by political-economic forces in influencing agrochemical use. In particular, researchers have revealed that marketing policies of unscrupulous, profit-seeking transnational corporations that manufacture pesticides (Weir and Schapiro 1981), lack of effective regulatory policies in developing countries (Bottrell 1984; Goldberg 1985), class relations in agriculture (Wright 1986), constraints imposed by poverty (Bull 1982), the changing requirements of commercial agriculture for both external and internal markets (Goldberg 1985; Goldman 1986; Thrupp 1994), and control by capital and the state over the labor process in contract farming (Shipton 1985; Murray 1994) all contribute to growing dependency on and misuse of pesticides.

The focus on political-economic forces has certainly enhanced our understanding of pesticide-related problems. Such a perspective is essential for uncovering pressures leading to increased pesticide dependence and misuse. Such analyses are generally weak at the local level, however, because they fail to explore adequately the manner in which farmers' attitudes, perceptions, and strategies limit or exacerbate pesticide misuse (Goldman 1986).

In many studies that do emphasize behavior at the local level, the role of political economy is often ignored, and the tendency is to account for environmental problems by portraying farmers as being "careless" and "indiscriminate" in their use of pesticides. For example, Medina (1987: 157) asserts: "There is much overuse, misuse and abuse of pesticides by vegetable farmers in the Philippine Cordillera. Spraying pesticides has now become a habit, rather than a necessity, and spraying is done indiscriminately." Other researchers have also used the terms "careless" and "indiscriminate" in characterizing pesticide use in developing countries (see Zaidi 1984; Thaman 1985; Black, Jonglaekha, and Thanormthin 1987; Guan-Soon and Seng-Hock 1987; Mohan 1987; Stonich 1993). Certainly, indiscriminate and careless use of pesticides exists, and it aggravates problems associated with increasing agrochemical dependency. But researchers who emphasize only pesticide misuse create the impression that carelessness and indiscriminate use are characteristic of the rural communities they examine. They rarely emphasize or even consider the extent to which farmers are cautious in relation to pesticide use, though work by Goldman (1986) is a notable exception.

My own research, however, indicates that it is crucial to explore the dimensions of *both* caution and carelessness in pesticide use (Grossman 1992b). To examine these contrasting patterns, it is necessary to employ Johnson's (1972) concepts of "individuality" and "experimentation." "Individuality" is the tendency for villagers to make agricultural decisions based on their own individual preferences, experiences, needs, and perceptions, all of which vary from farmer to farmer. "Experimentation" is the tendency for farmers to experiment regularly with new cultivation techniques and crops, with the goal of improving their production potential. These processes, which are characteristic of small-scale farming generally, generate considerable diversity in farming practices among members of the same community. Although Johnson employed these concepts in the analysis of traditional agriculture, they are also highly relevant to understanding contemporary pesticide use; they contribute not only to problems of misuse but also help account for the coexistence of varying degrees of caution and carelessness in pesticide use in the same community. To illustrate these processes, it is necessary to examine pesticide use in both banana and food crop production, because they are closely linked.

Pesticide Use in St. Vincent

Villagers are quick to voice their mixed opinions on the subject of pesticides. Many feel that ground provisions used to taste sweeter before the advent of these agrochemicals. They also blame them for the decline in a variety of terrestrial fauna, such as snakes and manicou (a small possum hunted for food), and crayfish, which a few capture by pouring the pesticide Sevin into streams. Others worry about the effects on their own health. Although opinions vary on the numerous impacts of increased pesticide use, most concur that it is also an integral element in contemporary Vincentian agriculture.

Pesticide use in St. Vincent had its origins in the era before World War II. Farmers, primarily large-scale growers, applied lead arsenate and bordeaux mixture to control pests in arrowroot and cotton. Annual reports of the 1930s and 1940s, however, indicated that severe pest outbreaks were relatively rare. The only exception was in the fledgling banana industry of the 1930s, which suffered devastating blows from Panama disease.

Local food crops had even fewer pest problems, though they certainly did exist. For example, attacks by the ever hungry mole cricket (Gryllotalpidae), which inflicts damage by cutting off tender seedlings as it burrows slightly below the soil surface, affected tomatoes, cabbage, and other "greens." But the extent of problems was much less severe in food crops then compared with today. Peasants used a variety of traditional methods to limit the depredations of insects (see Collymore 1984). Some sprayed soapy water or a mixture of pepper and water on their crops. Lighting fires at night with the hope that insects would fly into the flames was another method. Burning fallow vegetation before planting killed insect pests. And farmers simply handpicked them off their crops.

The growth of pesticide use in St. Vincent, as elsewhere in developing countries, is primarily a post–World War II phenomenon. The organochlorines—most of which are now banned for use in U.S. agriculture because of their persistence in the environment and their ability to accumulate in the food chain—initially became the pesticides of choice. Thus, DDT replaced lead arsenate in arrowroot production, and banana growers in the revived industry started to use aldrin and dieldrin in the late 1950s and 1960s and later heptachlor to control the banana borer that was affecting the industry. The chemical cornucopia that was slowly invading the land was reflected in a 1957 Department of Agriculture booklet entitled "How to Care Your Budded Citrus Plants," which recommended the use of aldrin, dieldrin, lead arsenate spray, chlordane, and parathion, all of which are banned for use in U.S. agriculture today, except parathion, which is extremely toxic to humans and the use of which is highly restricted. Although precise quantitative data on the extent of pest problems in arrowroot, cotton, and food crops before and after World War II are not available, the impression gained from reading numerous Agricultural Department reports is that the severity of attacks increased in the early postwar era. The precise cause of the growth of pest problems in both local food crops and export crops has not been established but is likely the result of: growth in pest resistance to persistent organochlorine pesticides, a widespread phenomenon (Bull 1982; Wright 1986); the decreasing length of fallow periods; and the relatively long periods that bananas occupy the same plots. Pest problems have continued to mount in Vincentian agriculture over the last twenty to thirty years, in spite of the eventual move away from the organochlorine pesticides. One female farmer's observation about increased pest problems and decreased effectiveness of pesticides captured the essence of this trend: "Worms [pests] know all the tricks now."

Dependence on pesticides has grown considerably in St. Vincent, with the value of such imports increasing from only EC$21,041 in 1955 to EC$7,417,000 in 1993 (SVG 1956; SVG Statistical Unit 1995). In 1989, approximately eighty-two pesticide formulations were available for purchase on the island (Isaacs

1989). The banana industry clearly consumes the largest amount; for example, in 1988, the SVBGA imported over 90 percent of all bulk pesticides brought into the island nation (SVBGA files; SVG Statistical Unit files), selling them to farmers from its large warehouse located in Kingstown.

Farmers also apply pesticides on a variety of food crops, such as cabbage, tomatoes, and carrots, with cabbage and tomatoes receiving the highest doses. Peasants purchase some pesticides for use on food crops at several small retail outlets in Kingstown and apply others obtained from the SVBGA, though the Association imports such agrochemicals with the expectation that they will be applied only to bananas.

Characteristic of many developing countries (Bottrell 1984), St. Vincent has no comprehensive and effective system for regulating the importation, sale, and use of pesticides. Legislation enacted in 1973 (SVG 1973) created the Pesticide Control Board to regulate pesticide-related matters, but it was vague concerning the board's powers. De facto control has been exercised by the Ministry of Agriculture, Industry, and Labour, which has been more restrictive than most government agencies in other developing nations in controlling pesticide importations. Only one of the major pesticides used in St. Vincent in the last few years, the herbicide Pilarxone, is considered inadequately tested or banned for use in industrialized countries.[1] The Ministry of Agriculture, Industry, and Labour currently prohibits the importation of several pesticides, including DDT, aldrin, dieldrin, heptachlor, DBCP, toxaphene, aldicarb, and parathion, all of which can be found in other developing countries, even though they are banned or highly restricted for use in the United States (Bull 1982; Zaidi 1984; Goldman 1986). Of all the pesticides widely used in St. Vincent, only one, Gramoxone (paraquat), is on the Pesticide Action Network's "Dirty Dozen" list.[2]

Banana growers obtain "official" information about pesticides from the SVBGA. The Association, in turn, traditionally received recommendations from WINBAN, which tested and evaluated agrochemicals on its own research farm in St. Lucia, charging the agrochemical manufacturers for all costs associated with its field trials. Although Vincentians do not use the most worrisome pesticides, banana growers nevertheless apply some that are quite toxic. In particular, nematicides used in banana production—Mocap, Furadan, Vydate, and Miral, which kill microscopic nematodes in the soil that attack banana roots—are all highly toxic if ingested, and the foul-smelling Mocap can be dangerous when in contact with bare skin. Furthermore, the recently introduced organophosphate nematicide, Miral (isazofos), which has become quite popular among banana farmers in St. Vincent in the mid-1990s, was linked in newspaper reports in the second half of 1995 to two deaths and several other cases of poisoning symptoms in farmers in three separate incidents elsewhere in developing countries.

The herbicide Gramoxone, the most widely used pesticide on the island, is also very toxic and has received much negative publicity because ingestion of the chemical is a major means of suicide not only in St. Vincent but in many other developing countries (Wright 1990). These nematicides (except Miral) and Gramoxone are classified as being for "restricted use" in the United States; because they can pose a significant threat to human health and the environment, only specially licensed and trained workers are permitted to purchase and apply them.[3] In contrast, any farmer in St. Vincent can buy and use them. Pesticides intended specifically for food crops, however, such as Sevin and Malathion, are less toxic, though their misuse can still create health and environmental problems.

The Political-Economic Context of Pesticide Use in Banana Production

Application of the majority of pesticides on export crops is a pattern widespread in developing countries (Goldberg 1985), and the case of St. Vincent is no exception. In the prewar era, growers applied most agrochemicals to cotton and arrowroot. Since then, developments in the banana industry have been primarily responsible for the nature and growth of pesticide use on the island. Agrochemical application in the industry is a reflection of a development model that continues to focus on export agriculture (see also Wright 1990) and of pressures from capital and the state to improve farm productivity and fruit quality.

In the early stages of the industry in the 1950s, losses due to pests were minimal. But as farmers continued to replant bananas in the same fields, pest attacks multiplied, a condition facilitated by planting densities of 650 to 700 per acre.

These pest problems, coupled with farmers applying less fertilizer than officials recommended, contributed to low yields. Consequently, banana officials in the Windwards industry and their advisers in the BDD, the British agency responsible for designing and administering British foreign aid programs in the region, decided that it was necessary for peasants to apply more agrochemicals—both pesticides and fertilizers—to boost productivity per unit area. They perceived the potential for improved yields as being considerable, viewing with envy the impressive banana yields (up to twenty tons per acre per year) achieved in the nearby French-speaking islands of Martinique and Guadeloupe in the 1960s, where farmers were applying much higher levels of chemical inputs than were Windwards growers (Phillips and Twyford 1965: 14). In contrast, at the time, Windwards farmers obtained relatively poor yields, three to four tons per acre.

The problem of low fruit quality was also depressing farmers' incomes. High-quality fruit has a variety of characteristics, including being fully formed and well shaped, characteristics dependent on healthy root systems, adequate nutrient intake, and proper growth rates. Because pests in St. Vincent and elsewhere in the Windwards were interfering with these conditions, they were a constraint on improving fruit quality and hence profitability.

To encourage banana growers to use more agrochemicals, the British government, through its aid programs, supplied the SVBGA and the other three Windward Islands banana growers' associations with free and subsidized pesticides and fertilizers on numerous occasions. These programs also supported several banana replanting and rehabilitation schemes from the late 1960s to the early 1980s. These schemes provided subsidies for chemical inputs and financial support for more intensive banana agricultural extension efforts, both of which facilitated greater use of pesticides and fertilizers.

Indeed, foreign aid is a classic channel through which advanced capitalist countries foster intensive agrochemical use in developing nations and create overseas markets for their domestic corporations. Thus, the dominance in St. Vincent and elsewhere in the Windwards of the herbicide Gramoxone, a product of the British firm Imperial Chemical Industries, reflects in part its liberal distribution through British foreign aid programs in the 1970s. Another example illustrates the key role that foreign aid can play. In the late 1980s, the Ministry of Agriculture, Industry, and Labour was initially hesitant to allow importation of the paraquat-based herbicide Pilarxone from Taiwan because it did not have enough data on its safety and effectiveness. Nonetheless, political pressure from the Taiwanese, who provide aid for several agricultural projects on the island, resulted in approval of the product in the early 1990s. The SVBGA subsequently sold it to growers but in 1994 stopped offering the herbicide for sale when it found that it was too potent and was, in the words of one SVBGA official, "killing everything."

The SVBGA also attempts to stimulate agrochemical use through the provision of institutional credit, a classic mechanism in contract-farming schemes for such purposes. A long-standing policy of the SVBGA has been to grant farmers who purchase inputs interest-free credit with liberal repayment schedules. Furthermore, SVBGA extension services—as in other contract-farming schemes—are an important source of influence in motivating peasants to intensify pesticide use.

The unscrupulous marketing practices of transnational corporations selling pesticides in developing countries is a much discussed topic in the literature (see Weir and Schapiro 1981). In the banana industry, their marketing has been done primarily through WINBAN and the growers' associations, and WINBAN's testing of most (but not all) pesticides used in the industry acted as a buffer against the more severe disasters reported elsewhere. Transnational corporate contact with peasants is limited mostly to catchy radio advertising, although very infrequently company representatives also make presentations before small groups of banana farmers, touting the benefits of their products.

Nonetheless, pesticide companies can contribute to problems of misuse. The manufacturer of the nematicide Mocap provided the Windwards banana growers' associations with backpack applicators to enable farmers to apply precise dosages. But an investigation in St. Lucia in 1989 revealed that the company had calibrated the machines to deliver forty grams of Mocap per banana plant instead of the thirty suggested by WINBAN, which, of course, resulted in heavier than recommended dosages and higher sales for the company (Bousquet 1989: 1). Concern about dosages of Mocap was initially raised in St. Lucia when "farmers began complaining about unusual dizziness while applying it" (ibid.). My own subsequent investigation testing two Mocap applicators in St. Vincent revealed dosages approximately 10 to 20 percent higher than those recommended by WINBAN. Another example concerns the same nematicide: Mocap is also used in the United States, but the bags sold in St. Vincent carry the label "For Export Only," thus raising the possibility that different formulations—of unknown consequence—are used for the export market. Certainly, problems were not limited to companies selling pesticides for bananas. Perhaps the most blatant disregard for human safety was an advertisement aired on television in which a man was happily spraying his house with an insecticide-filled aerosol can to get rid of insects that were bothering him; he even sprayed his pillow and a few seconds later could be seen sleeping contentedly in his bed—with his head on the same pillow!

Food production for sale in local markets has also become increasingly dependent on pesticides, though to a much lesser extent than in the banana industry. Government pressure on farmers to use pesticides on food crops is minimal compared with that in banana production, in part because pest problems are less severe on food crops and also because, historically, government support for agriculture in St. Vincent and elsewhere in the Windwards has usually focused on exports crops. Farmers' contact with government agricultural extension officers, who provide advice on pesticide use on food crops, is also much less frequent than it is with SVBGA extension agents. Furthermore, government involvement in regulating the sale of local food crops is marginal.

Nevertheless, growth in chemical use in banana production has influenced trends in local food crop cultivation. Although British aid programs and SVBGA credit schemes have been aimed specifically at banana cultivation, they have also stimulated increasing agrochemical use in food crop production; many farmers apply herbicides and insecticides obtained from the SVBGA on their local food crops.

Chemical dependence in local food cultivation also parallels that in the banana industry in its reflection of the changing requirements of commercial agriculture. Similar to the problem faced by banana growers in their overseas market, Vincentian farmers producing for sale in local markets are finding their customers increasingly selective concerning the quality of produce that they are willing to purchase. Undoubtedly, consumer exposure to the image of perfect, blemish-free produce through viewing United States–based television and through overseas migration has influenced the trend toward increasing selectivity.[4] Blemishes that were once ignored by customers ten or twenty years ago make produce unsalable today. Farmers realize that tomatoes marked with black spots caused by flying insects they call "booboo" or cabbages riddled with numerous small cavities caused by the diamondback moth (*Plutella xylostella*) and the budworm will not be marketable. Consequently, Vincentians strive to produce blemish-free crops for the local market by using more pesticides.

Patterns of Pesticide Use in Restin Hill

Pesticide use has become an integral part of Restin Hill agriculture. Ninety-seven percent of the households involved in agriculture had used at least one synthetic pesticide during the period July–August 1988 to July–August 1989, with the average being four (see Table 7.1).[5] Such widespread adoption of pesticides is characteristic of most Vincentian communities. Collymore (1984: 215), who interviewed a sample of farmers in the northeast and northwest parts of the island in the early 1980s, found 94 percent of those surveyed used pesticides (see also Wedderburn 1995).

The most widely used pesticide in Restin Hill and elsewhere in St. Vincent is the herbicide Gramoxone (paraquat) (see Table 7.2), which farmers apply in banana fields and, to a lesser extent, in the production of food crops. In an agricultural system in which labor shortages are a major constraint, Gramoxone is particularly appealing. The next most widely used pesticide is Primicid, which the SVBGA imports to sell to banana growers to control the banana borer, but farmers also apply it on their food crops. Villagers use the nematicides Furadan and Mocap mostly in banana cultivation, whereas Basudin and Sevin are the most frequently used agrochemicals applied exclusively on nonbanana crops.

TABLE 7.1. Distribution of Number of Pesticides Reported Used in Restin Hill, by Household, from July–August 1988 to July–August 1989

Number of Pesticides Reported Used	Number (%) of Households
0	1 (3)
1	0 (0)
2	3 (8)
3	10 (27)
4	11 (30)
5	6 (16)
6	4 (11)
7	1 (3)
8	1 (3)

Sources: Interviews with thirty-seven households; Grossman 1992b.

Whereas pesticide use in banana cultivation is widespread, it is more variable in food crop production (Table 7.3). Villagers in Restin Hill do not apply pesticides on their root crops and only very infrequently on some others, such as beans, because pests rarely affect their marketability.[6] But almost all farmers apply insecticides on the two most economically important garden "greens," tomatoes and cabbage.

As is true in other developing countries, pesticide misuse certainly occurs in Restin Hill (Grossman 1992a). Although the SVBGA urges farmers to wear protective clothing, such as boots, gloves, and masks, when applying pesticides, most do not, which can lead to excessive exposure through inhalation and skin contact; for example, those spraying Gramoxone in their fields can occasionally be seen wearing shorts or going barefoot. Inhaling the mist from Gramoxone spray, which can damage lung tissue, is made more likely by the windy conditions that prevail. Even more dangerous is application of granular nematicides—Mocap and Furadan—without proper protective clothing. In a standard reference manual on pesticides, *MSDS Reference for Crop Protection Chemicals* (1989), the manufacturers of both products warn that inhaling the products can be fatal. In the case of Mocap, the fine dust from the granular nematicide is easily dispersed into the air and can be inhaled and collect on applicators' clothing.[7] A small number of farmers dip seedlings of garden greens in liquid solutions of pesticides using their bare hands. Similarly, I was informed that a few farmers apply Mocap with their bare hands, which is harmful because it can be absorbed through the skin.

Lack of protection when applying agrochemicals reflects several constraints, including the fact that farmers find the protective clothing too uncomfortable to

TABLE 7.2. Number of Households in Restin Hill in Which Members Reported Using Various Pesticides, from July–August 1988 to July–August 1989

Pesticide Brand Name	Common Name	Number (%) of Households
Gramoxone	Paraquat	35 (95)
Primicid	Pirimiphos ethyl	33 (89)
Mocap	Ethoprop	19 (51)
Sevin	Carbaryl	19 (51)
Furadan	Carbofuran	14 (38)
Basudin	Diazinon	13 (35)
Other[a]	—	9 (24)
Decis	Deltamethrin	3 (8)
Malathion	Malathion	3 (8)

Sources: Interviews with thirty-seven households; Grossman 1992b.
[a]"Other" includes cases in which farmers did not know the names of the chemicals and cases in which they used spray cans containing insecticides intended for residential use on their food crops.

wear in the hot, humid weather typical of the area. An incident involving a representative of a transnational corporation speaking to a group of banana farmers about the benefits of his company's nematicide is illustrative; he first displayed pictures showing proper protective clothing—a man covered from head to foot in a protective suit and wearing gloves and a respirator mask—which immediately drew laughter from some in the crowd, with one amused farmer declaring that the protective suit was likely to kill him in the hot weather before the pesticide did! In addition, farmers' misunderstanding about the health impacts of pesticides are relevant to their limited use of protective clothing. Many villagers mistakenly associate how dangerous a pesticide is for their health with the strength of its odor. Similarly, many feel that pesticides will not harm them if they either eat a full meal or drink sugar water before applying them.

Methods used for measuring dosages can lead to applications that are either too strong or too diluted. Some determine the amount of water and pesticide to mix together based on the resulting solution's color or odor. Villagers also employ homemade measuring devices, such as bottle caps of varying types attached to the end of a wire or stick—which may fail to provide the precise amounts needed.

Generalizations about dosages that farmers apply are difficult to make. Most apply nematicides less frequently and in lower amounts than recommended. In contrast, some use concentrations of Gramoxone that are too strong, a problem

TABLE 7.3. Pesticide Use on Selected Food Crops in Restin Hill, from July– August 1988 to July–August 1989

Food Crop	Number (%) of Households Growing Crop	Number (%) of Households Growing Crop that Applied at Least One Pesticide on Crop
Dasheen	36 (97)	0 (0)
Tomatoes	29 (78)	27 (93)
Cabbage	26 (70)	25 (96)
Beans	26 (70)	3 (12)
Cucumber	26 (70)	8 (31)
Carrots	19 (51)	12 (63)
Sweet potato	16 (43)	0 (0)

Sources: Interviews with thirty-seven households; Grossman 1992b.

reported elsewhere in the Windwards (Rainey 1985); frustrated farmers increase concentrations of the herbicide beyond recommended levels when they fail to obtain satisfactory weed control, which can result from uneven spraying or from the invasion of Gramoxone-resistant weeds.[8]

Improper disposal of pesticide containers is another hazard. Empty bags that once contained toxic nematicides are supposed to be burned or buried, but most banana farmers let the bags rot in their fields, either placing them on cut banana pseudostems or discarding them on the ground. Some growers dispose of empty plastic Gramoxone bottles simply by throwing them away in their fields, whereas a small number, primarily those with low incomes, use them to store water for drinking or washing dishes or clothes.[9]

Caution and Variability in Pesticide Use

Certainly, such pesticide-use problems are widespread in developing countries, but researchers tend to present misleading images by focusing only on the dimension of misuse. In fact, farmers in Restin Hill are generally more cautious in their use of agrochemicals than would be expected from reading some of the literature on pesticide use in developing countries (Grossman 1992b). Indeed, Restin Hill farmers refer to pesticides as "poisons," which indicates their awareness of the possible dangers associated with chemical use in agriculture. Several practices reveal the extent of their caution.

The spatial organization of crops is one indication. When cultivating food crops separately from bananas in steeply sloping fields, farmers plant their food crops on the top portion of the hills and the bananas below. People fear that if bananas were planted above the food crops, runoff during rainfall would carry pesticides used in banana production downhill and contaminate their food crops.

Another indication of caution is their refusal to follow recommendations by extension agents from both the SVBGA and the government. The SVBGA tells banana growers to apply nematicides at the time of planting and then regularly at four- to six-month intervals, depending on the brand used. When villagers intercrop bananas with food crops, some do apply nematicides to bananas at the time of planting but do not reapply them until nine months later, after which they will have harvested almost all their food crops; others simply wait to make the first application nine months after planting. Such deviation from official practices results from villagers' fears that the nematicides would contaminate their food crops. Also, most banana farmers in Restin Hill previously refused to use Nemagon (DBCP), the nematicide that WINBAN recommended in the 1960s and 1970s, because they feared that injecting the foul-smelling liquid into the soil would contaminate it and their food crops; such hesitancy was fortunate because subsequent research has revealed that DBCP, which is now banned for use in St. Vincent, can cause sterility in humans (Thrupp 1988). Similar evidence of caution is evident in relation to food crop cultivation. The large majority of farmers will not spray cabbage with pesticides after the leaves have folded (which occurs about three weeks before harvest) because they worry that later applications would leave unhealthy residues on the crop, even though government agricultural extension officers reassure them that Decis can be sprayed safely on cabbage up to seven days before harvest.

Other evidence of concern about pesticide contamination exists. Many farmers who have cut bananas for home consumption (because the bunch is not suitable for export) will not leave them to ripen on the ground in their banana fields because they believe that pesticide residues there would contaminate the ripening bunch.

Caution is also revealed when examining the age and sex distribution of pesticide users. Realizing that these chemicals can be harmful, parents do not allow their children to apply them. The youngest to use pesticides are fifteen-year-old males, who sometimes apply Gramoxone, though the task is usually reserved for older individuals. Because people believe that men in general are more resistant than women to the dangers from pesticides, women are much more hesitant to apply them, and some women, especially those who are pregnant, refuse to use them at all. Data from the formal interview on pesticide use that I conducted are illustrative of the predominance of male involvement in pesticide handling. Thirty-two of the thirty-seven households in that sample contained both male and female adults (fifteen years of age and older), enabling women to rely on men to apply agrochemicals. In such cases, women prefer to

TABLE 7.4. Female Use of Pesticides in Households That Included Adult Males and Adult Females, Restin Hill, from July–August 1988 to July–August 1989

Pesticide	Number (%) of Households Applying the Pesticide[a]	Number (%) of These Households in Which Women Applied the Pesticide
Nematicides	25 (78)	6 (24)
Gramoxone	31 (97)	13 (42)
Primicid	29 (91)	12 (41)
Sevin	17 (53)	12 (71)
Basudin	11 (34)	3 (27)

Sources: Interviews with thirty-seven households; Grossman 1992b.
[a]Thirty-two of the thirty-seven households had both adult males and females.

let men (either their husbands, cohabiting male companions, or adult male children) apply nematicides, the pesticides that villagers fear the most. Thus, in those households that did apply nematicides, women did so in only 24 percent of the cases, whereas men used them in 88 percent (Table 7.4). For two other widely used pesticides, Gramoxone and Primicid, which people fear less than the nematicides, women used them in only 42 percent and 41 percent of the households applying them, respectively. The figure for women using Gramoxone reflects, in addition to health concerns, the heavy weight of full knapsack sprayers (over fifty pounds). The highest proportion of female use of a pesticide is 71 percent, which is for the insecticide Sevin and which people (correctly) believe is much less dangerous for them than most of the other agrochemicals used regularly.

Cautious use of pesticides also reflects, in part, economic considerations. They are expensive, and careless application would be economically wasteful. Such concern is evident in hiring agricultural wage laborers. Forty-six (seventeen out of thirty-seven) percent of the households in this sample had hired farm workers in the previous twelve months, but only five had hired laborers to apply pesticides, mainly for Gramoxone spraying in banana production. The limited hiring of workers for pesticide-related tasks reflects farmers' concerns about uncaring, careless workers possibly wasting expensive chemicals and also contaminating their food crops.

The evidence thus indicates that a certain degree of caution exists with regard to pesticide use; we clearly cannot generalize that Vincentian farmers are careless and indiscriminate. Nevertheless, as indicated, pesticide misuse in Restin Hill also occurs. But such misuse is not uniform. The key to understanding the issue of pesticide misuse lies in examining the extent and significance of variability in pesticide-related practices and beliefs. For example, whereas many are cautious when applying pesticides, others are not. Although some faithfully follow the instructions of SVBGA extension officers concerning proper chemical use, others do not. Some refuse to use any pesticides, as indicated by one woman who asserted, "They [pesticide users] may be gambling with their lives, but I'm not gambling with mine"; in contrast, a few farmers apply them with their bare hands. Practices and attitudes in relation to pesticides are highly variable. The following examples explore the extent of variability further.

Patterns of pesticide application are one arena of variation, as the case of cabbage cultivation indicates. Twenty-six of the thirty-seven respondents in the survey (70 percent) grew cabbage. Considerable diversity is evident in relation to pesticide use in nursery establishment, transplanting, and tending the growing plant. Twenty-three of the twenty-six applied pesticides when setting seeds in their cabbage nurseries to control attacks by ants, but the choice of agrochemical varied. Ten growers used Primicid, ten preferred Sevin, two relied on Furadan, and one used Mocap. At the next stage, transplanting from nursery to garden, the majority used Primicid, but methods of application differed; seven farmers dipped just the roots of seedlings into Primicid, six completely immersed them in the pesticide solution, two sprayed the furrow before transplanting, and one sprayed the seedlings after transplanting. Four others applied Basudin, and one put on Sevin. The general consensus was to apply pesticides after transplanting only when pest infestations occurred, but the choices varied—most preferred Basudin or Sevin, but others relied on Decis, Malathion, Furadan, or Primicid. Similar variability occurred in pesticide use on other crops, such as tomatoes.

A second example of differences among farmers concerns beliefs about precautions necessary to prevent pesticides from damaging people's health. The majority believe that dietary patterns affect people's ability to resist damage from pesticides. Remarks such as "If you are hungry, poison [pesticides] works through the body because you are empty," "Sugar water works out the poison," and "Resistance [to pesticides] is weak if you don't eat before" reflect local beliefs.

The specific nature and timing of the dietary precautions followed, however, varied. Fifteen farmers indicated that they ate a meal before applying pesticides, four combined a meal with drinking sugar-sweetened water beforehand, and eight preferred to consume a meal beforehand and drink sugar-sweetened water after using pesticides. The other responses included just drinking sugar-sweetened water, having some rum afterward, and not taking any dietary precautions. Although the prophylactic value of consuming such items is doubtful, the patterns are a further indication of individual variations.

The third example concerns application of the herbicide Gramoxone, used mainly in banana production. Although the majority of its users (64 percent) mixed Gramoxone only with water (which is environmentally the most sound practice), others sometimes added either kerosene or gasoline (or both) to the mixture. People used such additives either to provide a "sticking" agent (to help keep the sprayed Gramoxone attached to the plant leaf surface long enough to ensure desiccation) or to save money (the additives are less expensive than the herbicide). A few of those now using only Gramoxone had previously added these fuels but no longer do so because they either heard that the fuels could harm the soil or found them to be ineffective.

Individuality and Experimentation

Understanding the coexistence of caution and carelessness and the general pattern of variability evident in beliefs and practices is essential to uncovering the dynamic processes that contribute to pesticide misuse in contract farming. In particular, Johnson's (1972) analysis of individuality and experimentation as sources of innovation and change in traditional agriculture is highly relevant. Building on his contribution, I would add that the environmental rootedness of agriculture, which creates conditions of production that are inherently variable in space and time, encourages such individuality and experimentation. In such a context, a simplified, uniform set of production practices—similar to that sometimes found in mass production in industrial settings—would clearly be inappropriate for all farming conditions, and consequently farmers traditionally devise, on their own, particular solutions to their own unique circumstances—patterns reflected in individuality and experimentation.

In the contemporary scene, individuality and experimentation generate variability in pesticide-related behavior and attitudes in both local food crop production and banana production. Farmers experiment with pesticides to solve problems and improve yields. One example of experimentation that generates variability is the formulation of "pesticide cocktails"—that is, the combination of more than one pesticide in applications, which may be ineffective, wasteful, and in some cases hazardous to human health and the environment. The basis of such formulations is not official instructions but farmers' own initiatives. For example, Collymore (1984) revealed that cultivators in the northern part of St. Vincent devised on their own a variety of pesticide combinations to control the depredations of rats. I found evidence for the practice in Restin Hill: for example, the use of both Primicid and Basudin in the same application to control insect pests and the combining of kerosene or gasoline with Gramoxone for weed control. The application of such pesticide cocktails occurs elsewhere in the Caribbean (Pollard 1981; Murray and Hoppin 1992; Patterson 1996), as well as in other developing countries (Fagoonee 1987; Medina 1987; Murray 1994). Another example of experimentation is the varying dosages formulated in applying Gramoxone. Individuality is evident in the wide range of practices associated with the application of nematicides in banana production, from some refusing to use them at all to a few others applying them with their bare hands; such practices reflect attitudes ranging from intense fear of pesticides to minimal concern.

Although individuality and experimentation are characteristic of all farming communities, the significance of these patterns for pesticide use can vary in different contexts. Clearly, in some cases, peasants have limited autonomy in decision making with regard to pesticides, as can happen with tightly regulated contract-farming schemes; pressures to follow regimented pest-control practices may constrain, though not eliminate, variability generated by individuality and experimentation. Similarly, in some contract-farming schemes, the buyers/central coordinators provide their own pesticide-spraying crews to ensure control over the process (Murray and Hoppin 1992: 601). In other situations, however, as with the Vincentian case in which capital and the state have limited control over cultivation practices, farmers have greater autonomy in decision making concerning agrochemical use. In such cases, individuality and experimentation will produce considerable diversity in pesticide-use practices.

When pesticide misuse occurs, we cannot explain the exact nature of the problem by considering only political-economic forces, though their influence will be considerable. Similarly, at the local level, we cannot simply attribute the problem to "ignorance" or "lack of knowledge," although lack of knowledge may contribute to some problems (see Goldberg 1985; Medina 1987). Rather, problems of misuse also stem from the dynamic qualities of individuality and experimentation. Even when Restin Hill farmers are clearly aware of official recommendations concerning pesticide use, they often devise their own strategies. In essence, these farmers make decisions regarding agrochemicals based not on ignorance but on their combined consideration of market pressures, official recommendations, their own experiments, advice from other farmers, and their own experiences, preferences, and assessments of risks. Such complex influences clearly distinguish the nature of the labor process in agriculture from that in assembly line manufacturing involving deskilled laborers.

Certainly, economic factors, such as income and amounts of land controlled, are likely to explain some variability in pesticide use among farmers. Although I did not collect data on incomes or areas cultivated for all thirty-seven households in the pesticide-use sample, it is possible that such economic variables influenced some patterns, such as the number of pesticides used or amounts of money spent on these agrochemicals. Similarly, low incomes can influence pesticide problems when some farmers are too poor to purchase protective equipment. Most of the variability described in this chapter, however, is independent of such economic variables and is reflective of individuality and experimentation.

Individuality and experimentation, by leading to a variety of patterns of pesticide use, tend to produce problems of misuse. Consequently, farmers use pesticides in numerous situations unintended by the officials importing them. The case of Primicid, imported by the SVBGA, is a prime example. The SVBGA and the Ministry of Agriculture, Industry, and Labour recommend that Primicid be used only in banana cultivation (to control the banana borer). But villagers, on their own initiative, also apply it on their food crops. They use it at the time of setting cabbage and tomato seeds; at transplanting (usually dipping either the roots or entire seedlings in a solution containing the liquid); and, less frequently, after transplanting. Not only is Primicid supposed to be used only in banana cultivation, but it is also supposed to be applied only on the soil and not directly on plants. Farmers, however, do not use the same-strength solution of Primicid in food crop production as they do in banana cultivation, believing that such a strong mixture would contaminate their food crops and inhibit plant growth. Instead, they use much weaker solutions, which they determine by the color of the liquid mixture, by measurement, or by smell (several say they mix just enough Primicid to "catch the scent"). Thus, they not only have devised uses for Primicid unintended by the SVBGA but also have devised their own dosages. The ultimate impacts of such actions are unknown; although Primicid can leave residues on crops, the extent of the residue problem in St. Vincent is unknown because the manufacturer's original testing of the product was based on recommended solutions, not diluted ones. Another possible problem is that regular use of such diluted mixtures may actually facilitate pests developing resistance to Primicid. Similarly, a few farmers apply Furadan inappropriately in cabbage cultivation. The SVBGA imports Furadan to enable farmers to control nematodes that attack bananas. Use of Furadan on crops such as cabbage, however, may be hazardous because the active ingredient in Furadan (carbofuran) can be translocated and stored in leafy green matter, which, in the case of cabbage, people eat. The pattern of pesticides imported for use on one crop being subsequently applied on others can be found elsewhere in the Caribbean (Pollard 1981; Rainey 1985).

Because experimentation and individuality combine to produce variations in pesticide-use patterns, it is highly likely that some of these variations will be safe for people and the environment, whereas others will be harmful. The problem for farmers is the difficulty in evaluating the comprehensive, complex impacts of agrochemical use. Although they can assess the immediate, short-term effects on controlling pests on particular crops, analysis in relation to longer-term health and environmental impacts is much more difficult. Whereas in traditional agriculture farmers could readily evaluate the results of experimenting with new crop varieties or cultivation techniques, in contemporary agriculture their understanding of the implications of their diverse pesticide-use practices for such environmental processes as resistance to pesticides or biological magnification becomes much more problematic. Similarly, assessing pesticide impacts on human health is difficult because many of the symptoms of low-level pesticide poisoning, such as headaches, nausea, and dizziness, can be attributed to a wide range of illnesses.

Interestingly, the variability evident in patterns of pesticide use reflects the failure to deskill labor. If the SVBGA had more effective control over the labor process, it could prevent such variation. And certainly, the Restin Hill case is not unique. Instances have been reported elsewhere in which limited corporate control over peasant contract farmers has led to higher than acceptable levels of pesticide residues on crops (Murray 1994). In essence, farmers are not as easily controlled as industrial workers, and hence greater variability in production patterns—in pesticide use and in other aspects of farming—is to be expected.

Conclusion

Contract-farming schemes have profound significance for peasantries in developing countries. While providing them with a guaranteed market, capital and the state also attempt to intervene into and shape the peasant production process. Moreover, as this study indicates, the level of technological innovation in contract farming often surpasses that in previous forms of agriculture, and the degree of technological complexity can increase over time. That contract farming has significant impacts is without doubt. The key issue here is the appropriate framework for analyzing this novel institution.

Certainly, the growth of contract farming can be linked to patterns described in the literature as being Fordist and post-Fordist. In relation to Fordism, such schemes can provide fresh foods for mass diets and standardized inputs of uniform quality for food-processing industries. In relation to post-Fordist trends, contracting can be an essential element in providing capital with flexible sources of supply in an era of vertical disintegration, fragmenting markets, and strategies of global sourcing. But in relation to the Windward Islands banana industry, such frameworks provide, at best, limited benefits. While it is possible to argue that bananas were a standardized commodity destined for mass markets in the United Kingdom, the conceptual seeds of contract farming in the industry were established by the Imperial Economic Committee (1926) over seventy years ago, before Fordism became established in the United Kingdom. And although the literature on post-Fordist trends emphasizes the importance of competition based partly on quality, such a consideration has long been a key aspect of competition in banana markets; for example, Geest was already stressing in the 1950s that "improved quality was the only means whereby producers could meet such [Jamaican] competition successfully" (*West India Committee Circular* 1956a: 101). Thus, as others argue in relation to patterns of industrial change (see Page and Walker 1994), the periodization offered by the literature on regulation theory and flexible specialization does not fit well with the historical patterns revealed in the history of the banana industry in the Windward Islands.

Even more problematic is the portrayal of agriculture, in general, and contract farming, in particular, as being either Fordist or post-Fordist in relation to the labor process (see also Goodman and Watts 1994; Page and Walker 1994). As the case of the Windward Islands indicates, characterization of the labor process within such frameworks presents special difficulties. Certainly, we cannot consider banana farmers as deskilled laborers performing repetitive tasks, as is supposedly typical of Fordist industry. Indeed, peasant contract farming, by its very nature, requires a coupling of conception and execution. Nor is it appropriate to consider such growers as the enskilled, multitask workers described in the literature on flexible specialization as being characteristic of a segment of the contemporary industrial workforce. Although the skill content in banana farming has increased over time, such change has not been a function of the pressures leading to similar patterns among industrial workers described in the literature on flexible specialization. In the banana industry, the growing skill content of labor has not been an abrupt adjustment in response to capital's attempts to cope with increasing competition and rapid market changes since the mid-1970s. Rather, it has been a slow, evolving pattern that has been ongoing for almost fifty years, as both capital and the state have shifted previously off-farm processes onto the farmers. Moreover, although both a core of the contemporary industrial labor force and banana contract farmers in the Windwards are "multitask" workers, peasants have always been multitasking—producing and integrating the cultivation of local, traditional food crops, export commodities, tree crops, and livestock. The basic problem in applying perspectives developed to analyze changes in industry to the realm of agriculture and contract farming is that the labor processes in industry and agriculture are fundamentally different, a function of the environmental rootedness of farming.

The literature on the globalization of agriculture, represented by work on "the new international division of labor in agriculture" and on the "new internationalization of agriculture" does provide certain crucial insights into changes in contract farming in the contemporary era, while also exhibiting certain weaknesses. A major theme in this literature—as well as that on globalization generally—is the decline in the regulatory power of the nation-state and the rise of supranational forces. Most clearly, the decline in the ability of the British government to make its own decisions regarding support and protection for the Windward Islands banana industry has been a function of the growing power of the EC and its successor, the EU and its Single European Market. Such changes have had significant, negative impacts on contract farming in the Windwards: increased quotas for Latin American fruit, decreasing prices for Windward Islands bananas, and intensification of the labor process (in response to new EU regulations concerning the packing of bananas). And such pressures have intensified because of supranational regulatory institutions; the GATT and its successor, the WTO, have helped spur trade liberalization, and the recent WTO ruling against the EU's banana import regime could have devastating consequences for the Windwards.

But what is particularly interesting about the process of globalization is not simply its homogenizing tendencies concerning the decline in the power of the state but how states mediate such forces and the implications of that mediation for the local level (Whatmore 1994). In essence, globalization and homogenization inevitably meet locality. As Raynolds et al. (1993: 1104) declare, "Individual states, of course, have differing capacities to negotiate this process." And the Windwards will certainly have limited ability to negotiate this process, given their weak economic and geopolitical influence; indeed, they have always had very limited control over the external events and processes that have long influenced the islands during both colonial and postcolonial periods (Richardson 1992), thus calling into question an impression created by the globalization literature that the weakness of nation-states is a contemporary phenomenon. Nevertheless, even Vincentian state policy, in responding to the forces of globalization, will still have significant impacts on the lives of banana growers. For example, the policies of the SVBGA, in coping with these pressures, have led to a further intensification of the labor process, letting the farmers experience the full brunt of the impacts. But such a result was not inevitable. The SVBGA could also have removed from the farmers the increased burdens associated with cluster packing and the new EU regulations by reestablishing the system of boxing plants used in the 1970s and first half of the 1980s, where SVBGA personnel could—at centralized locations throughout the island—cluster, wash, treat, and pack the fruit (a recommendation that I have made to the SVBGA), thereby alleviating the unbearable labor burdens on the growers.

Mittelman (1996b: 229) observed that globalization "must eventually touch down in distinct places." The manifestations of the processes of globalization reflect local political-economic, cultural, historical, and environmental forces that must be the subject of investigation. Two examples are relevant here. One relates to the assertion in the literature that technologies of production have become homogenized, reflecting their crucial significance for global sourcing (see Sanderson 1986b). Yet the case of the Windwards demonstrates clearly that nonstandardized—and rapidly changing—technologies (related to packing bananas) can be employed that are adapted to local conditions but are still oriented toward producing standardized output. The technologies used in the Windwards reflect adaptations to the constraints imposed by the evolution of land-use patterns, current inequalities in control over land, impoverishment and exploitation of the industry by Geest, the islands' steep slopes, and fragmentation of holdings. The other example concerns food production. Although forces associated with globalization can undoubtedly contribute to declines in local food production, such results are not inevitable. Moreover, when food production does decline, it is not necessarily related to globalization, as the extensive literature on the impact of colonialism indicates (see Watts 1983; Grossman 1984a). Furthermore, when forces associated with globalization are involved, they are usually not the only factors contributing to the problem; they interact in complex ways with other political-economic and environmental forces, and such patterns require investigation. Political ecology provides an appropriate framework for the analysis of such issues and contract farming generally, because it is sensitive to the interaction among local and global forces, political economy, and the environment.

Political Ecology and the Environment

The field of political ecology developed initially to provide a political-economic perspective on the causes of environmental degradation. In the 1990s, research interests have broadened considerably (Peet and Watts 1996), but the concern with the fundamental importance of political economy remains central to the field. Similarly, in this book I have focused, to a considerable degree, on the significance of political-economic forces—in particular, the complex roles of the British state, capital (Geest), the Vincentian state and its statutory corporation regulating the banana industry (the SVBGA), the EU, GATT, the WTO, conflicts between British and U.S. interests, the evolution of differential control over land in St. Vincent, and the importance of wage labor relationships at the village level. But political ecology is more than just a detailing of political-economic influences and their relations to human-environment interactions. Rather, the purpose is to illuminate issues of significance in the development literature, three of which are highlighted in this study—the labor question, the food question, and the environmental question.

While political ecology's strength has been in the analysis of political-economic forces, it can also be faulted for failing to give proper consideration to the significance of the environment itself (Grossman 1993). Put simply, political-ecological relations are not "environment free." My concern with the importance of the environment is certainly not unprecedented. Traditional cultural-ecological studies paid much more attention to the role of environmental variables than political ecologists do today (see Hardesty 1977; Netting 1986). My argument is not that the environment is more important in our explanations than political economy but that the former has not received the attention that is warranted in political ecology. In particular, the concern should not be just how political economy affects the environment, precisely because the relationship is not unilinear. Rather, we need to examine how political-economic and environmental variables interact to shape human-environment relations.

It is worthwhile to consider Zimmerer's (1996b: 179) evaluation of political ecology and his suggestion for future research: "By way of conclusion we might note that its future contribution could show a considerable success if the nature of environmental modification is more fully and recursively integrated with theories of regional development and underdevelopment." I agree with his assessment but would add that we need to expand beyond the issue of integrating environmental modification into our analyses. Environmental modification should be viewed as only part of a broader interest in the importance of the environment itself. In particular, the significance of the environment for understanding human-environment relations is based on two dimensions. First, environmental variables are creative forces that help shape resource-use patterns and interact in complex ways with the influence of capital and the state. Second, the environmental rootedness of the farming experience encourages growers to develop more flexible labor patterns than industrial workers. Both these dimensions have important implications for the analysis of the labor question, the food question, and the environmental question.

The Labor Question

The central issue of the "labor question" in contract farming revolves around the implications of control by capital and the state over the peasant production process. That a degree of control exists is clear—but how thorough it is and the significance of such control are the subject of debate.

Looked at from the vantage of what is written in contracts and what official policy pronouncements of the contracting agencies are, the appearance of rigid control is evident. But practice is often more complicated than pronouncements. Indeed, it is appropriate for this study to alter Clapp's (1994: 81) statement concerning contract farmers that "the farmer's control is legal but illusory" to assert that "control by capital and the state is legal but often illusory." Certainly, we cannot generalize about the degree of control. And attempts to develop a model of a generic form of contract farming based on such rigid and overwhelming control (compare Clapp 1994) masks more than it illuminates (Little 1994). A political-ecological perspective helps situate the issue of control, as it focuses on both political-economic and environmental forces.

Certainly, a degree of control is evident in the case of banana contract farming. Capital and the state have been able to introduce a series of increasingly complex technologies aimed primarily at the stage of harvesting and packing. And such introductions have had substantive implications for the labor process as well as food production. But effective control has been more difficult to achieve in the other stages of production.

Control in contract farming rests, in part, on the regular and intensive provision of extension services to educate and monitor farmers. British American Tobacco's scheme in Kenya is illustrative: "BAT's strategy has hinged on extension. . . . BAT established a ratio of one front-line extension worker per 50 farmers—a flood of extension agents, by African standards. . . . BAT supervised its agents very closely, making them enter their remarks in notebooks on the farm, which their supervisors could check. The agents visited every farm fortnightly, like it or not. . . . Technically speaking, it worked" (Shipton 1985: 293–94). But compared to this situation, the provision of extension services has been inadequate to control banana growers' cultivation practices in the Windward Islands. Such inadequate supervision is also reflected in the inability of the SVBGA to determine the precise number of growers involved in the industry. The weak extension service is even more surprising given the technological complexity of banana contract farming and the rapid pace of technological change.

Understanding the poor state of the extension service requires consideration of capital-state relations and, more recently, EU-state relations. Given Geest's reaping of the majority of revenues from the banana industry, the SVBGA and the other Windward Islands banana growers' associations never had sufficient funds to provide adequate extension services. A considerable amount of British aid for the numerous replanting and rehabilitation programs was directed at improving these services, but it was never enough to make up for the shortfalls. In the current context, the new policies of the EU and the SEM are continuing to drain the growers' associations financially, perpetuating the inability of the Windwards' extension services to achieve effective control over labor.

But environmental forces have also played a fundamental role in contributing to the financial weakness of the SVBGA. The perpetual crises caused by incessant environmental disasters have helped cripple the finances of the associations. Indeed, this contribution to weakness in extension services and the resulting inability to control the labor process in cultivation is another example of how environmental and political-economic forces interact to shape resource use.

Another instrument of control in contract farming is the imposition of quality standards and grading specifications. But establishing such criteria does not necessarily guarantee complete compliance on the part of the peasantry. In the Windward Islands banana industry, the specification of quality standards has been most effective in achieving control in the harvesting/packing stage, especially since the introduction of field packing. Unless growers deliver fruit meeting specifications and packed according to directives, the SVBGA will downgrade or reject the bananas. But while specifications appear rigid, complete compliance by peasants is not achieved, as is evidenced by continuing quality problems.

But extension service supervision and grading standards are not the sole contributors to compliance with regulations. In the case of the banana industry—and certainly in other contract-farming schemes—the degree of correspondence between official policy and farmer implementation has just as much to do with banana prices as it does with control by capital and the state over the peasant labor process. Thus, low prices limit the degree to which peasants follow the directives of capital and the state. For example, minutes of an SVBGA meeting (SVBGA 1983: 2) reported that "the low banana prices in 1982 accounted for low production in 1983 since many farmers, discouraged, neglected field care, husbandry, sanitation and even harvesting." Similarly, cultivation practices in banana production have deteriorated significantly in the 1990s in response to poor returns. In contrast, when prices are comparatively high, as they were in the latter half of the 1980s, villagers are more likely to follow official procedures. Consequently, an analysis of banana contract farming in the late 1980s alone would have led one to assume that capital and the state had considerable control over the labor process, but in reality, such compliance at the village level was, to a significant degree, really a function of what villagers felt were remunerative prices. Thus, researchers focusing on contract farming must be more sensitive to the role of prices in facilitating compliance with regulations, a consideration that has significant implications for determining the degree of control by capital and the state.

This discussion also suggests that it is inappropriate to consider "control" in general terms, as is usually done in the literature. Capital and the state rarely try to control the entire labor process, instead seeking to control discrete fragments of the process, while leaving other aspects to the farmers. For example, they usually do not attempt to control labor recruitment strategies, leaving that domain—with its social costs and potential conflicts (see Carney 1988)—to the contract farmers. Moreover, control is never static. Such selective and changing control has been evident in the banana industry. Control by capital and the state over peasant labor has increased over time, but primarily in the stage of harvesting and packing—not in other aspects of banana production.

Control must also be understood in both historical and cultural contexts, which further suggests that discussions of control in contract farming in generic terms is inappropriate. For example, Carney (1988, 1994) has shown how the intensive labor requirements in contract farming of rice in The Gambia led to gender-based, intrahousehold struggles over land and labor that were fought in the context of cultural idioms, conflicts that ultimately thwarted the ability of the state to control the labor process (see also von Bülow and Sørensen 1993). Moreover, cultural and historical contexts affect the susceptibility of farmers to control. In the case of the Windward Islands, resistance to estate domination and control during slavery and the postemancipation era has been an enduring feature, as it has elsewhere in the English-speaking Caribbean (Mintz 1985, 1989; Richardson 1992). This long-standing tradition of resistance conditions and limits the degree of docility and malleability of the contemporary peasantry in the face of demands by capital and the state. Nor are there traditions of rigid labor control within households or communities. Male control over female labor is often problematic, especially as many women control their own property and obtain varying degrees of financial independence through marketing activities (Momsen 1993). Furthermore, whereas strong, unilineal descent group structures in some communities in other developing countries, such as those in Africa, may culturally condition members to a certain level of control from outside their households, communities in the Windwards lack such groups, a function of the social disruption and forced relocation associated with slavery.

Discussion of the issue of control also leads to a consideration of the appropriateness of describing contract farmers as "disguised wage laborers." The concept clearly implies thorough domination by capital and the state, which is not applicable to the case of the Windward Islands banana industry and its contract farmers. Moreover, these growers and other contract farmers are clearly differentiated from wage laborers in other ways—they accumulate capital, reinvest income from banana sales in land purchases, hire laborers (and thus extract surplus value), and creatively manage the labor of family, relatives, and friends in the context of changing technological requirements (see also Little 1994: 225–27). Moreover, some banana contract farmers have exercised an "exit option" (Hyden 1980), decreasing—and in some cases withdrawing completely from—cultivation of bananas and increasing their production of food crops in the 1970s and again in the 1990s in response to unremunerative prices, a shift made possible by their control over the means of production—land.

Another crucial issue in the "labor question" that is related to the issue of control is Braverman's concept of the deskilling of labor. The three dimensions of the concept—the separation of conception and execution, the fragmentation of the labor process, and capital's domination of labor—are clearly inadequate to capture the dynamics of the labor process in banana contract farming. Having already discussed the third dimension, we can focus on the first two.

Contract farming in the banana industry involves a melding of conception and execution, not a separation. Although official technologies are devised by capital and the state, farmers implement them in a variety of ways as they attempt to cope with labor constraints, their unique environmental resources, and variable prices. Moreover, they have to integrate food crops, bananas, tree crops, and livestock spatially and temporally, which requires creativity on their part. Peasants must make decisions throughout the cultivation cycle—in selecting planting materials, forms of tillage, spacing of plants, follower setting, method of propping, and patterns and timing of fertilizer and pesticide applications—decisions affected, in part, by the peculiarities of their particular plots—their relative location, slope, size, aspect, likely rainfall, soil type, drainage, exposure, and unique crop histories. Farmers must integrate all these considerations in an attempt to maximize output from their small plots of both bananas and food crops, at the same time striving to produce fruit that meets increasingly rigid quality specifications. In essence, farming is not a monotonous, assembly line process with a simple, standardized script to follow outlined by the dictates of capital and the state. The environmental rootedness of agriculture requires farmers to meld conception and execution.

Such melding also makes control by capital and the state over the labor process more difficult in peasant agriculture than in industry. Indeed, effective control implies a uniformity of production practices. In contrast, this study has documented a wide range of variability in such activities. Certainly, effective control would also lead to consistently high levels of quality output, a goal that the Windwards have failed to achieve.

Similarly, deskilling's notion of the increasing fragmentation and simplification of the labor process is not applicable to the Windward Islands banana industry. In fact, the opposite pattern has occurred as tasks performed previously off-farm have been shifted by capital and the state onto the farmers—making the labor process more complex over time and requiring increasing levels of skill in banana production. As others have noted (Wood and Kelly 1982; Grint 1991), what is crucial for capital is not the simplification of the labor process per se but the amount of surplus that can be drained from the labor process. Geest profited by shifting tasks that it once performed—dehanding, deflowering, clustering, and packing—to the Windward Islands. In turn, the SVBGA shifted tasks it once performed—dehanding, deflowering, fungicide application, packing, and weighing—to the growers in an attempt to reduce its own costs in light of its perpetually weak financial condition and hopefully to improve the quality of fruit exported from the island. And as the labor process became more complex, growers had to readjust their labor recruitment strategies to cope with intensified packing and harvesting demands coupled with declining prices—adjustments that are an integral part of the conception process in contract farming.

The basic question of whether this Windwards case is really unique, an exception to the "normal" conditions of control in contract farming, must be raised. A review of the literature reveals that the degree of control evident in this case is certainly not an isolated occurrence. Several examples indicate the various reasons for failure to achieve thorough control. First, as noted previously, studies indicate that intrahousehold, gender-based struggles can lead to problems of labor availability to carry out the dictates of contracts (see Carney 1988; von Bülow and Sørensen 1993). Second, inadequate extension supervision is certainly not limited to the case of the Windwards. Laramee (1975) describes a Thai onion contracting scheme in which farmers failed to follow the specifications of contracts, in part because of weak extension supervision. In addition, the Thai farmers gave preference to their own crops, to the consternation of the contracting agency; he (ibid.: 50) notes that "[t]he best planting days came and went and no contract crops were sown." Furthermore, farmers' activities reflected their own understanding of the peculiarities of their environments, just as in the Vincentian case. Laramee (ibid.) reports, "The farmers had their own experience and preferences of many years telling them just the opposite of what the Company was advising them to do." And Glover and Kusterer (1990: 127), in a review of contract-farming schemes, observe: "Companies also complain that growers do not follow their technical advice or suggested planting dates. Growers often try to economize by applying less than the recommended quantity of chemicals and they may plant when it is convenient, rather than on the date agreed." Similarly, in an overview of contracting in sub-Saharan Africa, Little and Watts (1994: 17) assert, "Where labor is not tightly regulated and controlled, growers have been able to bend rules and manipulate regulations to their own advantage, while at the same time minimizing the negative effects of contract farming." A problem in some of the literature is that the image of contract farming that has been created has been shaped primarily by what Watts (1994a: 64) terms "[a]uthoritarian and despotic forms of contracting"—those in which control is very effectively achieved. In reality, contracting exhibits wide variations in place and time.

The Food Question

The literature on the globalization of agriculture links the decline of food production and the growth of food imports in developing countries to the process of globalization. In many cases, trade liberalization, structural adjustment programs, and policies of transnational corporations facilitate such patterns. But an exclusive appeal to such general homogenizing tendencies fails to provide an adequate understanding of national and regional agricultural trajectories. Political ecology, with its focus on both political-economic and environmental forces, is particularly suited to the analysis of this issue, given its attention to the interplay between local contexts and broad, structural forces.

A key issue in the literature is the relationship between contract farming for export and declining local food production and food import dependency. In the case of St. Vincent, conflicts in the productive sphere related to control by capital and the state have not been the cause of the decline in local food production, contrary to the conventional wisdom. Banana planting and harvesting occur year-round, and the most intensive labor demands for bananas and for food crops fall on different days of the week. Although bananas are labor intensive, the ability to cope with resulting labor shortages is more a reflection of marked inequalities in control over land and resources, not of conflicts in the productive sphere. Moreover, as bananas and food crops are usually planted in the same fields in a variety of patterns, spatial conflicts are also not of prime significance, except in the case of cattle raising.

But control by capital and the state has affected food production in a more novel manner. The mandatory adoption of increasingly complex forms of harvesting and packing has made banana production more immune from theft compared with the situation in food crop cultivation, and given the seriousness of the perennial problem of praedial larceny, such consequences have placed food production at a relative disadvantage.

Blaming banana contract farming for the growing dependency on food imports also ignores crucial dimensions of the Vincentian context. The basic problem is that Vincentian farmers find it difficult to compete with low-cost food imports from subsidized, large-scale U.S. agribusiness—not because of banana production but because they are disadvantaged by several constraints: a traditional lack of government support and inadequate extension services for local food production; small and fragmented plots that increase the costs of marketing; low levels of technology; and plots in marginal environmental settings. These patterns characterize the English-speaking Caribbean (McIntosh and Manchew 1985; Axline 1986; Thomas 1988), reflecting not the homogenizing pressures of globalization but the historical development of St. Vincent's political economy as well as its environmental context.

Researchers that blame the banana industry for declines in local food production fail to realize that Vincentian peasants sell most of the foods that they produce, and thus market conditions for local food crops—and not the nature of banana production—are critical determinants of the level of food production. Although markets for local food crops are plagued by problems typical of open markets, the banana industry, with its guaranteed market and regularity and promptness of payments, provided a more favorable situation for Vincentian peasants, a function of the pivotal support traditionally provided by the state—both British and Vincentian—for export agriculture.

Thus, a host of political-economic forces are relevant to comprehending the relationship between contract farming and local food production. But while the analysis of political economy is central to this study, environmental conditions have also been creative forces that need to be considered. The nature of St. Vincent's light soils encourages villagers to replant every two to three years, facilitating frequent intercropping. The steep slopes also facilitate intercropping because the system of tillage in food crop production in such an environment helps reduce runoff-induced loss of agrochemicals applied to bananas. The highly variable growth patterns of the banana plant, which is a perennial, make rigid control by capital and the state over the timing of activities in the phases of cultivation difficult to achieve; in contrast, in other systems of contract farming in which annual crops are involved that have more uniform patterns of growth than bananas, capital and the state can more effectively control the seasonal timing of labor inputs into the contracted crops.

Such patterns also indicate the importance of focusing on the complex interactions among environmental and political-economic forces. For example, unequal control over land has contradictory implications for the relation between food crop production and banana cultivation. The need to engage in wage labor by those with limited holdings creates labor shortages that negatively affect their involvement in both banana and food crop production. At the same time, such inequality leads peasants to cultivate small plots on steep slopes; gardening on small plots encourages intercropping, as peasants attempt to maximize output from their limited holdings, and cultivating on steep slopes encourages intercropping to limit the loss of agrochemicals due to runoff. And the nature of the particular type of banana chosen for export, the Cavendish group, has contributed to the endless round of technical innovations, as capital and the state have devised numerous practices designed to cope with the susceptibility of the Cavendish cultivars to bruising. Such innovations, in turn, have also influenced the relation between local food crops and bananas by intensifying labor shortages in agriculture generally and by making bananas less subject to theft. And how farmers cope with such labor shortages is conditioned by inequalities in control over wealth.

No inevitable association exits between contract farming for export and local food production. Control by capital and the state over the production process—a basic component of all contract-farming enterprises—will inevitably influence that relationship, just as it has done here. But the exact manifestation and significance of such control will be conditioned by a host of political-economic and environmental conditions that cannot be homogenized or reduced to a simple formula.

The Environmental Question

From the perspective of political ecology, pesticide misuse is of especial concern for several reasons. First, and most obvious, it involves the problem of environmental degradation. Second, a variety of political-economic forces attempt to shape agrochemical use. And, third, the analysis of pesticide use is highly relevant to the issue of the deskilling of labor and the distinctive nature of the labor process in agriculture. In particular, it also helps highlight the environmental rootedness of agriculture.

A variety of political-economic forces have shaped the growing emphasis on pesticide use in the banana industry. The perpetuation of St. Vincent's historic dependence on export agriculture and the resulting susceptibility to increasing demands for higher-quality fruit in the British market have been significant, a process that has intensified as a result of concentration in the quality-conscious food retailing sector there (Nurse and Sandiford 1995). Furthermore, British aid has encouraged and subsidized the use of agrochemicals generally, and the SVBGA's policy of liberal credit has also led to more intensive pesticide use.

But while these influences help explain growing dependency on agrochemicals in contract farming, they do not illuminate why pesticide misuse occurs. To understand patterns of misuse adequately, we have to focus more closely on the local level and, in particular, on *interpersonal* variations in perceptions and beliefs. Such an emphasis is consistent with recent trends, as noted by Blaikie (1995b: 204), who observed, "The social sciences . . . have significantly shifted towards a more interactionist mode in which there is not an objective reality, but many subjective ones which are provided by different people who see their 'real' landscape in their own ways." But such subjective differences in perceptions and their influences on behavior must not be viewed exclusively through the lens of groups defined by their political-economic characteristics—as is fashionable in political ecology today. Otherwise, we will fail to understand the dynamics of pesticide misuse.

Johnson's (1972) concepts of individuality and experimentation—which emphasize such interpersonal variations—have been particularly useful for analyzing pesticide misuse. They help explain the coexistence of caution, carelessness, and misuse in regard to agrochemical applications; indeed, what researchers often consider "carelessness" is really a function of experimentation, a prime example of which is the creation of pesticide cocktails. These characteristics—

which can be found in all farming communities to some degree—generate variations even among members within the same households. The key point is that individuality and experimentation have a dynamic of their own and lead to numerous forms of pesticide use, irrespective of the nature of intervention by capital and the state in the production process.

What is especially important is that individuality and experimentation are functions of the environmental rootedness of agriculture. Growers develop their own individual styles of farming in dealing with the inherent temporal and spatial variability of the environment. They also experiment to cope with such constraints and to improve productivity, quality, and output. Each farmer has a unique set of land-based resources with peculiar configurations of slope, aspect, drainage, altitude, soil type, and fertility that is affected in different ways by daily and seasonal changes in rainfall, temperature, wind, and pests. And given the political economy of land tenure on the island, the set of resources that farmers control will invariably change over time. The variability in farmers' experiences is compounded further because they plant a range of crop combinations, plant numerous varieties of the same crops, and employ different tillage techniques, which interact in complex ways with such environmental features. Such divergent environmental experiences help shape the nature of individuality and facilitate experimentation.

Given such patterns, the environmental rootedness of agriculture encourages farmers to develop their own flexible and changing strategies that are embodied in individuality and experimentation. In contrast to industrial settings in which technologies and resources can be homogenized globally, the natural resource base and its inherent variability present significant constraints to such global homogenization. In particular, the environmental rootedness of agriculture will ensure that "standardized" technologies in contract farming (see Sanderson 1986b), such as pesticides, will be employed in a variety of "nonstandard" ways.

Clearly, such divergent patterns of pesticide use are also relevant to the issue of deskilling, but in a novel manner. Assuming that capital and the state were able to devise appropriate pesticide-use programs (which is certainly not always the case), deskilling would result in more environmentally sound practices, as farmers—functioning in the context of the separation of conception and execution—would faithfully implement such directives. But pesticide misuse is perhaps one of the clearest—and most widespread—examples in contract farming of the failure to deskill labor. Indeed, individuality and experimentation, which are intimately linked to such problems, are integral elements of conception in agriculture.

The degree of individuality and experimentation will not be uniform in all contract-farming enterprises. Local and regional differences in cultural and historical backgrounds will affect such patterns. And in those cases in which capital and the state more effectively regulate and control production practices through more intensive agricultural extension efforts, these characteristics, which can be found in all farming communities, will be suppressed to some degree. But they will not be eliminated. Arising from the environmental rootedness of agriculture, they will persist as significant forces shaping contract farming and human-environment relationships.

But they also do not function in a vacuum. Political-economic forces remain fundamental influences on the labor process in contract-farming schemes. The essential task is to illuminate how environmental variables, individuality and experimentation, culture histories, and political economy all interact to shape human-environment relations. Political ecology provides a particularly powerful framework for such an endeavor. And central to such analyses must be a greater appreciation of the significance of the environmental rootedness of agriculture and its implications for understanding the uniqueness of the farming experience.

Notes

Introduction

1. The applicability of the term "peasant" to small-scale cultivators in the Caribbean has been the subject of much debate (e.g., Frucht 1967; Mintz 1973, 1983, 1985, 1989). I follow Trouillot (1988), who uses the term to refer to those involved in the peasant labor process. Focusing on the labor process rather than on the essence of a universal peasant type permits the inclusion of the numerous small-scale cultivators in the Caribbean who must rely on varying degrees of involvement in wage labor. Peasants produce both subsistence and cash crops, relying primarily on household labor, though they may both hire labor occasionally and supply it. They control the land they cultivate, although the degree of control can vary from ownership to sharecropping relationships. Peasants are concerned with both household reproduction and accumulation of capital. In the Vincentian context, most of those cultivating ten acres or less can be considered peasants.

2. Nontraditional exports are those that either have not been produced in a country or were not exported before (Thrupp 1994).

3. Some of the literature suggests that when both peasant and large-scale growers are involved in contract-farming schemes, contractors give preferences and benefits to the latter (e.g., Morvaridi 1995), but it is not possible as yet to make generalizations about the issue.

4. Many studies do show that contract farmers have higher incomes than do other growers (Minot 1986; Glover and Kusterer 1990), but exceptions exist (Little and Watts 1994).

5. Certainly not all sectors of the economy were characterized by Fordist techniques (Schoenberger 1988; Jessop 1992). Also, the nature of Fordism varied in North America and Europe according to differences in historical developments and sociopolitical structures (Roobeek 1987).

6. The flexible specialization approach distinguishes between mass production and flexible specialization or craft production as technological paradigms (see Piore and Sabel 1984). It claims that the crises of the 1970s have led to a qualitatively new era in which patterns associated with flexible specialization dominate.

7. As Schoenberger (1988) notes, standardization in Fordism did not mean com-

plete homogenization of product types, but the extent of product variability was limited owing to the nature of mass production techniques.

8. A related literature uses the terminology of regulation theory to examine different "food regimes" (see Freidmann 1991; Friedmann and McMichael 1989).

9. For example, states are forced to cut back on subsidies to domestically oriented agriculture, reduce barriers to the importation of food, and devote resources to the expansion of export agriculture, irrespective of the welfare or distributive implications of such policies (see Sanderson 1986b).

10. For reviews of the field, see Bryant 1992, Peet and Watts 1996, and Zimmerer 1996b.

11. Some researchers focusing on environmental regulation and homeostasis (e.g., Rappaport 1968) preferred not to use the term "cultural ecology," but such work is grouped together here for purposes of comparison (see Grossman 1977).

12. As many have noted, Wolf (1972) was the first to employ the term "political ecology," but his usage did not have the momentous impact that Blaikie and Brookfield's work has had.

13. The issue of plurality of perceptions and definitions of environmental problems was also highlighted by Blaikie and Brookfield (1987a).

14. Trouillot (1988) characterized banana growers on the Windward Island of Dominica in the same way, but he did not consider whether banana production was an example of contract farming.

15. The concept of disguised wage labor involves more than the issue of control. It also signifies that capital exploits farmers by extracting surplus value from them in a manner similar to that in the capital–wage labor relation. The issue was central to discussions about the articulation of modes of production perspective (see also Currie and Ray 1986).

16. See Samatar 1993 for an interesting discussion of the significance of labor control and exploitation of workers in banana plantations in Somalia.

17. A more recent trend in the labor literature linked to discussions of flexible specialization is the analysis of patterns of "enskilling" associated with the need for contemporary, flexible workers who can operate a variety of machines and take more initiative in the work process, a function of the context of rapid innovation and design changes (see Piore and Sabel 1984; Storper and Scott 1986; Kaplinsky 1988; Storper and Walker 1989).

18. Research on these topics has produced conflicting results, reflecting in part differences in the intensity of commodity production, the nature of the commodity, area devoted to the crops, size of landholdings, social relations of production, and boom and bust cycles (Grossman 1984a; von Braun and Kennedy 1986; Kennedy and Bouis 1993).

Chapter 1

1. The twelve-member EC became known as the EU in 1993.

2. The first shipment of Jamaican bananas reached Britain in 1895, but regular shipments did not begin until 1901 (Sealy and Hart 1984: 14).

3. In the 1920s and early 1930s, fledgling attempts had previously been made in St. Lucia and Dominica to export bananas to the British and North American markets, but they were short-lived. St. Vincent made several experimental shipments of bananas to the United Kingdom starting in 1903, but the effort also proved commercially unsuccessful (Fraser 1986: 191). Earlier, the West India Royal Commission (1897: 49) recommended that St. Vincent, along with Dominica, establish a banana trade with New York, but the Vincentian administration never implemented such a scheme.

4. The importance of the Lacatan in the Windwards industry has declined as other, shorter varieties of the Cavendish group, especially the "Robusta" cultivar, but also the "Valery" and "Grand Nain," have become more prominent, but these have the same susceptibility to bruising.

5. In the case of Jamaica, the British government stopped direct purchases in 1953.

6. The British officially designated "dollar countries" at that time as Bolivia, Canada, Colombia, Costa Rica, Cuba, the Dominican Republic, Ecuador, El Salvador, Guatemala, Haiti, the Republic of Honduras, Liberia, Mexico, Nicaragua, Panama, the Philippines, the United States, and Venezuela. Restrictions on imports from dollar countries involved not only bananas but also other commodities such as grapefruit, grapefruit juice, and orange juice (*West Indies Chronicle* 1972: 412).

7. For an excellent discussion of the history of Geest, see Trouillot 1988. The official Geest story can be found in Stemman 1986.

8. These topics are discussed more fully in the next chapter.

9. Where transnationals have considerable involvement in production there, plantations in excess of twelve thousand acres can be found (Jim Fitzpatrick and Associates 1990: 9).

10. This ratio was modified in subsequent years.

11. Since 1976, the value of the Eastern Caribbean dollar has been tied to the U.S. dollar at EC$2.70 = US$1.00. Previously, the EC dollar was tied to the British pound at EC$4.80 = £1.00.

12. Before 1967, the EC was known as the European Economic Community, or EEC.

13. Although the goal of the EC was to facilitate free trade, Article 115 of the Treaty of Rome enabled the EC Commission to grant members of the Community the right to limit imports from other member states if those goods originated outside the EC and such imports would violate quotas on such goods. Thus, countries with preferential quotas on bananas, such as France, could prevent other EC mem-

bers, such as West Germany, from importing dollar fruit and then shipping it to France.

14. Later, in the 1980s, that informal agreement distributing market share among them allocated approximately 40 percent to both Geest and Fyffes and 20 percent to Jamaica Producers.

15. Subsequent Lomé Conventions were II (1980–85), III (1985–90); and IV (1990–2000).

16. From 1985 to 1990, dollar producers supplied between 16 and 31 percent of the market.

17. The SEM was actually established on 1 January 1993.

18. As of 1993, the twelve members of the EU were Germany, France, Belgium, Luxembourg, Ireland, Denmark, Italy, Spain, Portugal, Greece, the Netherlands, and the United Kingdom.

19. Major Latin American banana suppliers to the EU are Costa Rica, Colombia, Panama, and Ecuador.

20. For a more thorough discussion of the intricacies of the new regulations, see Nurse and Sandiford 1995.

21. The ECU is the European Currency Unit, which is based on a basket of EU currencies, with one ECU worth approximately US$1.15 in mid-1997.

22. It has been widely reported in the United States press that Carl Lindner, the head of Chiquita Brands International, is a major contributor to both the Democratic and the Republican parties. Press accounts suggest that Lindner's generous contributions to the 1996 reelection efforts of the Democrats influenced President Clinton to lodge the complaint with the WTO, a potential scandal that is being called "bananagate."

23. See Wiley forthcoming for a discussion of the difficulties of diversification away from bananas in the Windwards.

Chapter 2

1. Hurricanes have sustained winds of seventy-four miles per hour and higher, whereas tropical storms are defined as having sustained winds of between thirty-nine and seventy-three miles per hour.

2. A "crown" refers to the tissue connecting a "hand" of bananas to the stalk of a bunch. A bunch in St. Vincent usually has between six and ten overlapping hands. Each hand has between twelve and twenty "fingers," or individual fruits. In ripe bananas, the crown appears as the very dark part of the hand where the fingers join together.

3. Carefully controlled conditions on the boats—temperatures kept between 55°

and 56°F—are necessary to prevent the fruit from ripening prematurely while on board.

4. Contracts were changed to fixed-price agreements in 1966–67 and in 1972–73 at the request of WINBAN, but they subsequently reverted back to their original formats.

5. The contractual change did result in a significant, but brief, jump in the price received by the banana growers' associations (WINBAN 1995: 2), but continuing price pressures in the SEM and the low value of the British pound muted the benefits. Had the contractual changes not been made, however, the price to the Windwards would have been even lower.

6. Historically, Windwards farmers have been in the least profitable part of the chain, receiving a small portion of the ultimate retail price of bananas. Various reports have indicated that the percentage accruing to growers has fluctuated significantly over time, ranging from 11.5 (Demacque 1983: 125) and "little more than ten percent" (Thomson 1987: 45) to 20.4 percent (Nurse and Sandiford 1995: 49). Most of the retail price goes to those involved in shipping, ripening, and retailing.

7. In 1995, quality assessments in the United Kingdom shifted to the analysis of green fruit instead of ripe fruit.

8. For example, I saw boxes of Windwards fruit being crushed while being loaded onto a ship chartered by Geest in St. Lucia in 1989.

9. Fyffes is no longer owned by Chiquita (formerly United Fruit), having been sold to Fruit Importers of Ireland in 1986.

10. Elsewhere (Grossman 1994) I explore in more detail the issue of British aid to the Vincentian banana industry.

11. See Welch 1996 for a discussion of the importance of banana growers' associations in the Eastern Caribbean.

12. This figure has to be accepted as an approximation because of multiple registrations.

13. This discussion concerns the SVBGA's services provided during the period 1988–95. In response to deteriorating financial conditions, it was subsequently planning to reduce the level of services provided to farmers.

14. The cess varied from EC$0.06 to EC$0.08 per pound of bananas sold over the period from 1988 to 1995.

15. IBDs were formerly "boxing plants" built in the late 1960s and early 1970s. As of late 1995, the SVBGA planned to close all IBDs.

16. Those elected to the board of directors tend to be medium- and large-scale growers; for example, banana incomes of five of the seven elected members of the board of directors in 1988 varied from EC$56,000 to EC$234,000 in 1988, far beyond the average incomes of peasants growers (SVBGA files).

17. Associations in the other Windward Islands were in similar financial conditions, with very substantial deficits.

18. One of the common complaints about the banana growers' associations in the Windwards generally is that they are, in fact, biased toward helping large farmers (see Thomson 1987: 34).

Chapter 3

1. The term "plantation" is used in much of the Caribbean literature. In St. Vincent, however, people generally refer to such large holdings as "estates," a convention that I follow. No official definition of estates exists, though reference to a minimum size of one hundred acres is sometimes seen. Yet because Vincentians apply more liberal interpretations, even properties as small as twenty or thirty acres are still referred to as "estates," especially when they were formerly part of larger estates.

2. Rubenstein (1995), who has conducted a long-term study of illegal marijuana cultivation on the leeward side of the island, asserts that the crop's economic importance is quite substantial and even rivals that of banana production.

3. Part of the difference between the data for 1946 and that for 1972 was the result of different definitions, but the trend of gradual decline is still valid.

4. Much funding for agricultural and agriculturally related projects (such as road building) comes from external donor sources.

5. Slaves, however, were forbidden to grow sugar cane.

6. One account indicates that the area of such yam pieces was "thirty to forty feet square" (Cameron 1967: 12).

7. Higman (1984: 184) reports that the cane-harvesting/processing season on one Vincentian estate lasted from December 1807 to late May 1808. For much of the region, the season lasted from January to June or July (183).

8. Estimates at the time indicated that the size of Crown lands was 25,400 acres, but it was likely closer to 37,000 acres (John 1974: 135).

9. Fraser (1986) notes that the West India Royal Commission's data on lands held under ten acres undercount the amount held by peasants by a few thousand acres, but the fundamental inequalities are still evident in the distribution of land at the time.

10. Additional acquisitions in the 1930s for land settlement brought the total purchased to over eighty-five hundred acres (John 1974: 118). More land, roughly forty-six hundred acres, was acquired between 1946 and 1961. The Vincentian administration, however, initially operated the two largest properties acquired after 1946 as large-scale government estates, but it did subsequently lease some of the land to small-scale farmers. These properties, along with land acquired by the government in the 1970s and 1980s, became the basis of the current land reform program initiated in the late 1980s.

11. Vincentians in the Grenadines also produced another type of cotton for export in the 1800s, but it was never a significant export crop for the colony.

Chapter 4

1. Restin Hill is a pseudonym.
2. Unless otherwise noted, the "ethnographic present" is 1988 to 1989.
3. I was aware of only one household involved in agriculture in Restin Hill in which both spouses farmed but never worked together.
4. In male-headed households, men generally own the dwellings. In the four cases in which the female common-law spouse owned the dwellings, villagers still considered the households to be headed by the male spouses because they were the major income earners and the dominant forces within their households. I followed the villagers' designations in such ambiguous cases, a strategy also employed by Rubenstein (1987: 285).
5. I selected a stratified, random sample consisting of twenty-four of the sixty-four village households. I first divided the sixty-four into male- and female-headed units. I then divided both groups into six equal strata based on size of households' landholdings and randomly chose two from each stratum to yield twelve male-headed and twelve female-headed households for the sample. I collected data on the sample's gardens and on patterns of time allocation, banana production and sales, marketing activities, diet, income, and wage labor relations from November 1988 to August 1989.
6. Villagers could recall only one instance when a person from Restin Hill lost a property because of default on a mortgage payment.
7. As Rubenstein (1987) notes, some nevertheless attempt to convert such property to their own holdings or sell it.
8. A few individuals applied a different interpretation to the "one-third share" system. They felt that it was more appropriate for the farmer to keep three-fourths of the proceeds—that is, three parts for the farmer, one part for the landowner.
9. Based on data from the sample of twenty-four households.
10. While women today clearly dominate the marketing of food crops, some question their degree of involvement in such activity during slavery (e.g., Mintz 1989: 216–17; Mintz and Price 1992), though others (e.g., Marshall 1991) assert women also dominated food marketing then as well.
11. In the Restin Hill case study, I could obtain accurate data on incomes for only sixteen of the twenty-four households in the sample.
12. In her study of a banana-producing village in St. Lucia, Barrow (1992) also found several patterns similar to those which I discuss in this chapter.

Chapter 5

1. This figure is derived from my observations of how long it took numerous farmers to perform the various tasks involved in banana production. It assumes that growers perform all tasks recommended by the SVBGA, though many try to reduce their labor expenditures in a variety of ways that are discussed in this chapter. It includes not only farming activities but also time spent selling bananas off-farm, occasional visits to the SVBGA main office in Kingstown, and time lost for rain delays while working in their banana plots. A "labor day" is five hours.

2. The "plant" crop is the first crop obtained from a plant; subsequent crops from the same plant are referred to as "ratoon" crops.

3. The banana does not have a real stem, which is why it is called a pseudostem.

4. One of the problems with sleeving is the disposal of the bags after use. Some end up clogging waterways. Most are simply discarded in banana fields, where the elements break them down into small fragments and they become incorporated into the soil, the consequences of which are unknown. Unlike in Latin America, few Vincentian farmers use bags treated with pesticides.

5. For many years previously, the SVBGA bought bananas regularly on Tuesdays and Wednesdays.

6. The plant crop yields most of its fruit from nine to fourteen months after planting, with a peak at approximately ten months. The first ratoon is ready for cutting at around fifteen months and runs to the twenty-first or twenty-second month, with a peak at sixteen months, and the second ratoon lasts roughly from twenty-two to twenty-nine or thirty months after planting, with its peak usually at twenty-six months (Polius and Laville 1988). A third ratoon, depending on the level of field care and the number of "followers," will often continue into production up to almost thirty-six months after planting.

7. Although some women harvest fruit, the task is performed mainly by men.

8. The term "drogh" is used to refer to the carrying of banana boxes on one's head. It likely stems from the word "drogher," which referred to small vessels carrying sugar and other items during slavery (see Carmichael 1833: 195).

9. Hours include only time spent in the field, not time spent traveling to and from the gardens.

10. A cardboard box weights approximately three pounds.

11. The quality of regular-sized fruit from St. Vincent improved only marginally in 1996 to 76 percent (WIBDECO 1997: 17) but increased to 82 percent in 1997.

12. The study is based on the percentage of total time available in a twelve-hour day, seven days a week. It includes time spent on all activities, ranging from banana production, wage labor, eating, and child care, to resting and socializing. These data mentioned here, which are based on data from the twenty-four households in the sample, do not include time spent on wage labor in other people's banana fields. The

time allocation study is based on a large number (4,336) of random, spot-check observations of activities of all sample household members from November 1988 to August 1989; I randomly chose between six and nine days each month to make the observations, making observations on the same day of the week only twice in a month after all seven days of the week had already been selected. Because of the demanding nature of such an activity, I divided the sample into two groups of twelve households each and observed, with the help of a field assistant, the time allocation patterns of each group on alternating survey days. On each day of the study, I randomly selected which hour of the day each of the twelve households would be observed, never observing more than one household in an hour. We recorded only the activity that each person of the household was engaged in at the moment we first observed them within the selected hour for that household. The percentage of time spent on an activity is determined by dividing the number of observations made of that activity by the total number of observations of all activities. The types of activities in which people engage are influenced greatly by the day of the week. Consequently, in those months in which the same day of the week was selected twice, I randomly eliminated for this analysis those observations made on one of the two days. For more information on the methodology, see Grossman 1984b, 1993.

13. An interesting example of less than adequate reciprocation can be found in a case in which two men were helping another in swap labor to till his field. I noticed that the heights and depths of the ridges and furrows in the garden were smaller than those in other fields that I examined. I wondered whether this was an adaptation to the particular soil type. My field assistant quickly corrected me, pointing out that the comparatively small sizes of the ridges and furrows were really the result of a lackluster effort on the part of the two providing swap labor.

14. In other villages in the Marriaqua Valley and elsewhere in St. Vincent where average holdings were larger than those in Restin Hill, growers with more than ten to fifteen acres of bananas tended to rely primarily on wage labor.

Chapter 6

1. Engledow's study was the basis for the assessment of agriculture by the West India Royal Commission appointed in 1938. Known as the Moyne Commission, it was sent by the British to investigate economic problems and social unrest in the Caribbean and make recommendations for improving conditions in the colonies.

2. Similarly, food items made up 35 percent of all imports from 1935 to 1944 (Jolly 1947: xi). Although some distortion is evident because of the war years, the relative importance of food imports as a percentage of all imports appears to be a longer-term phenomenon.

3. In 1988, that number was even higher, at sixty-six.

4. These data are from food frequency interviews I conducted with women of twenty of the twenty-four sample households over a ten-month period, from November 1988 to August 1989. On days on which I conducted the time allocation study, I also attempted to ask either the household head, in the case of female-headed households, or the wife or common-law spouse, in the case of male-headed households, to mention all foods that she had eaten the previous day. On such days, I tried to interview as many of the women as possible, but because women were hesitant to discuss their consumption patterns in public, I could not interview them all each time. I compiled a total of 316 records of the previous day's meals, which are representative of both the households and seasonality. The data are in the form of frequencies, not total amounts consumed. Although it would have been preferable to interview all household members, it was impossible to do so because of time constraints. Nonetheless, the diets of these women are fairly representative of what others in their households ate, though adult males tended to consume some purchased foods in the rum shops that these women did not. Four of the women in the sample were uncomfortable discussing their diets, and thus I terminated such inquiries with them (see Grossman forthcoming).

5. I am indebted to Dr. Keith Nurse at the University of the West Indies, Trinidad, for bringing this point to my attention.

6. Figure 6.2 also includes imports of frozen turkey, but the vast majority of poultry imports were in the form of chicken.

7. However, prestige is sometimes associated with the types of imported foods consumed.

8. Vincentians had grown a variety of bananas before the advent of the export industry.

9. Based on data from the sample of twenty-four households. The reason that the two sets of data combined are over 100 percent is because the area intercropped with bananas and food crops is counted in the calculations for both bananas and food crops.

10. Cassava (*Manihot esculenta*), another root crop, is rarely found in the village today, having been more important in the past. Its cultivation is widespread elsewhere in the Caribbean (see Berleant-Schiller and Pulsipher 1986).

11. The seasonal nature of tree crop harvesting does not significantly influence the distribution of time spent in agriculture. First, farmers devote only minimal amounts of time to tree care. Second, although harvests of individual tree species are highly seasonal, they do not occur at the same time. Tree harvesting occurs throughout the year, as the harvests from one species yield to the next in a series of successions.

12. Major weeds include water grass (*Commelina* sp.), corn grass (*Setaria barbata* or *Rottboellia cochinchinensis*), and white head or white top (*Parthenium hysterophorus*).

13. This pattern helps to explain the uncharacteristically low level of intercropping found during my garden survey conducted from late 1988 to 1989. Many farmers had replanted their fields at the same time after a tropical storm hit St. Vincent and the Marriaqua Valley in September 1987. Thus, many of the provisions planted in that cycle had been harvested before my survey.

14. From August 1988 to August 1989, the number of sellers on Saturdays varied seasonally from 245 to 500 and on Fridays from 195 to 415. Market activity was slower Mondays through Thursdays, consisting of about fifty hucksters per day.

15. The long-standing market building was torn down in 1995 to make way for a more modern structure to house marketers.

16. The government does place price controls on local meats sold through retail outlets and on local fish sold in the Kingstown fish market, but not on food crops.

17. When some other crops, such as green or ripe bananas, are unsold, people give them away or simply leave them at the market at day's end.

18. Care must be exercised in interpreting the data in Figure 6.5. Earlier official reports on the production of provisions were based on government figures for provision exports. But these export data, particularly for the period 1984–86, were highly flawed because of the complicated currency transactions of traffickers. The data in Figure 6.5 are revisions of these earlier reports and are based on the 1985–86 national agricultural census, which also had problems. Nevertheless, the general overall trend in provision production revealed in Figure 6.5, if not the actual amounts, is accurate.

19. Data on exports in Figure 6.6 are based on records from the SVG Statistical Office, which must be accepted with great caution. Again, the trends reported are more accurate than the actual figures. Data on provision production after 1989 from the Ministry of Agriculture, Industry, and Labour were not available.

20. I developed these estimates with the help of Sylvester Vanloo of the SVBGA. Trouillot (1988) also reports significant consumption of bananas in Dominica.

21. Because the SVBGA has the exclusive legal power to export bananas, traffickers must get a license from the Association for all their banana exports.

22. The data in Figure 6.7 are based on harvests of fifteen of the twenty-four sample households. Information on harvests for these households from January to August 1989 came from field interviews and were corroborated by records of sales in the files of the SVBGA, whereas data on harvests from September to December 1989 (after fieldwork was finished) were obtained from SVBGA records on a subsequent visit to the island in 1990. No data were available for the other nine households because five did not grow bananas commercially, and the selling patterns of the other four (such as occasionally combining harvests with other households or selling under different names) made it impossible to determine how much they sold after I left St. Vincent in August 1989. Also, differences in amounts harvested per month, revealed in Figure 6.7, reflect, in part, the number of harvest days (Tuesdays and Wednesdays) that happened to fall in that month. For example, August had ten

harvest days, whereas July and September each had only eight, thus partly accounting for the peak in August.

23. Similarly, Hanseen (1990: 111), discussing marketing in Dominica, reports that "bananas are grown extensively by peasants and are intercropped almost all over the island." And Trouillot (1988: 135), in his study of banana growers on Dominica, observed that farmers often intercropped bananas with root crops.

24. Although it may seem that food crops would always benefit from fertilizer carried by runoff, such is not the case. Applying fertilizer to dasheen close to harvest time initiates chemical changes in the plants that make them less palatable.

25. Similarly, Welch (1996: 109) reports that bananas replaced sugar cane in St. Lucia.

Chapter 7

1. Granular Furadan, which is used in the banana industry, is not banned by the U.S. Environmental Protection Agency but is banned for use in the State of Virginia because it has been blamed for killing birds. Use of the granular product is gradually being phased out in the United States. Also, the Taiwanese agricultural aid mission has imported for its trials small amounts of pesticides whose formulation cannot be translated by the Plant Protection Unit.

2. The Pesticide Action Network established the "Dirty Dozen" list in 1985 to encourage stricter regulation of particularly hazardous pesticides that are sold worldwide but are banned, unregistered, or severely restricted for use in many industrialized countries. The list currently includes eighteen pesticides, of which nine are still used in the United States. Of all the pesticides on the list, paraquat (Gramoxone) is the least regulated worldwide.

3. Miral is not sold in the United States, but another pesticide, Triumph, which contains the same active ingredient, is; it also is classified for "restricted use." I could not obtain information on the exact differences between the two products, even after contacting the company that manufactures them both, Ciba Geigy.

4. This suggests interesting parallels with findings of research on the globalization of agriculture concerning the homogenization of quality standards (see Sanderson 1986b).

5. I used three methods to obtain information on practices and attitudes concerning pesticides. One was direct observation of activities. Another was informal discussions on the topic with numerous farmers. The third was formal interviews conducted at the end of the research period in July–August 1989 with farmers from thirty-seven of the fifty-two households engaged in agricultural production (cultivating at least 0.1 acres). My original goal was to complete interviews with one member from each of the farming households, but time constraints prevented me

from doing so. Nevertheless, the study includes a sample of sufficient size to be representative of the variability among pesticide users in the village. Members from both male-headed and female-headed households participated in the survey, with 62 percent of the respondents being from male-headed households. Those participating in the interviews, who were adults ranging in age from twenty-three to seventy-two years, were familiar with the pesticide-use practices of other members of their households. Landholdings of the households ranged from 0.1 to 15.6 acres. I conducted all the interviews.

6. In drier areas of the island, some farmers apply nematicides on sweet potatoes, but those in Restin Hill do not.

7. During the last few years, the manufacturer of Mocap has supplied a disposable mask and pair of thin, plastic gloves with each sack purchased. The glove and mask are supposed to be thrown away after one use, however, but given the small size of many of their plots, farmers do not use the entire contents of a bag at one time.

8. Villagers note that the weed called white head or white top (*Parthenium hysterophorus*) in banana fields is more common today, having become somewhat resistant to Gramoxone.

9. Many also use such empty containers to store water that will be used later to mix with pesticides.

Bibliography

Ahmad, Nazeer. 1987. *Land Capability of Saint Vincent and the Grenadines*. Washington, D.C.: Organization of American States.

Alexander-Louis, Theresa. 1992. *Economic Evaluation of Mini–Wet Cluster Method of Harvesting and Packing Bananas in St. Lucia*. Castries, St. Lucia: WINBAN.

Amin, Ash. 1994. *Post-Fordism: A Reader*. Oxford: Blackwell Publishers.

Amin, Ash, and K. Robins. 1990. "The Re-emergence of Regional Economies? The Mythical Geography of Flexible Accumulation." *Environment and Planning D* 8:7–34.

Amin, Ash, and Nigel Thrift, eds. 1994. *Globalization, Institutions, and Regional Development in Europe*. Oxford: Oxford University Press.

Arthur D. Little. 1995. *Study of the Impact of the Banana CMO on the Banana Industry in the European Union: Summary*. Paris: Arthur D. Little International.

Attewell, Paul. 1987. "The Deskilling Controversy." *Work and Occupations* 14:323–46.

Axline, W. Andrew. 1986. *Agricultural Policy and Collective Self-Reliance in the Caribbean*. Boulder, Colo.: Westview.

Ayako, A. B., L. M. Awiti, D. M. Makanda, Okech-Owiti, and G. M. Mwabu. 1989. "Contract Farming and Outgrower Schemes in Kenya: Case Studies." *East Africa Economic Review*, Special Issue (August): 4–14.

Barbier, E. 1989. "Cash Crops, Food Crops, and Sustainability: The Case of Indonesia." *World Development* 17:879–95.

Barkin, David. 1985. "Global Proletarianization." In Stephen E. Sanderson, ed., *The Americas in the New International Division of Labor*, 26–45. New York: Holmes and Meier.

Barrow, Christine. 1992. *Family Land and Development in St. Lucia*. Cave Hill, Barbados: Institute of Social and Economic Research.

Barry, Tom, Beth Wood, and Deb Preusch. 1984. *The Other Side of Paradise: Foreign Control in the Caribbean*. New York: Grove Press.

Bassett, Thomas J. 1988a. "The Political Ecology of Peasant-Herder Conflicts in the Northern Ivory Coast." *Annals of the Association of American Geographers* 78:453–72.

——. 1988b. "Breaking up the Bottlenecks in Food-Crop and Cotton Cultivation in Northern Côte d'Ivoire." *Africa* 58:147–73.

——. 1994. "Hired Herders and Herd Management in Fulani Pastoralism (Northern Côte d'Ivoire)." *Cahiers d'Études Africaines* 34:147–73.

Baxter, Vern, and Susan Mann. 1992. "The Survival and Revival of Non-Wage Labour in a Global Economy." *Sociologia Ruralis* 32:231–47.

BDD. 1975. "Windward Islands Banana Industry Improvement Programme." Report, document no. PC (75) 21, British Development Division in the Caribbean, Bridgetown, Barbados.

Beard, J. S. 1947. "Forestry." In Bernard Gibbs, ed., *A Plan of Development for the Colony of St. Vincent, Windward Islands, British West Indies*, 295–322. Port-of-Spain, Trinidad: Guardian Commercial Printers.

Beaver, Patrick. 1976. *Yes! We Have Some: The Story of Fyffes*. Welwyn, United Kingdom: Broadwater Press.

Beckford, George L. 1966. "Issues in the Windward-Jamaica Banana War." *Flambeau* 4:19–23.

——. 1967. *The West Indian Banana Industry*. Mona, Jamaica: Institute of Social and Economic Research.

——. 1972. *Persistent Poverty: Underdevelopment in Plantation Economies of the Third World*. New York: Oxford University Press.

Berleant-Schiller, Riva, and Lydia M. Pulsipher. 1986. "Subsistence Cultivation in the Caribbean." *Nieuwe West-Indische Gids* 60:1–40.

Bernstein, Henry. 1979. "African Peasantries: A Theoretical Framework." *Journal of Peasant Studies* 6:421–43.

Besson, Jean. 1987. "A Paradox in Caribbean Attitudes to Land." In Jean Besson and Janet Momsen, eds., *Land and Development in the Caribbean*, 13–45. London: Macmillan.

——. 1988. "Agrarian Relations and Perceptions of Land in a Jamaican Peasant Village." In John S. Brierley and Hymie Rubenstein, eds., *Small Farming and Peasant Resources in the Caribbean*, 39–62. Winnipeg: University of Manitoba.

Besson, Jean, and Janet Momsen, eds. 1987. *Land and Development in the Caribbean*. London: Macmillan.

Birdsey, Richard A., Peter L. Weaver, and Calvin F. Nichols. 1986. "The Forest Resources of St. Vincent, West Indies." United States Department of Agriculture, Research Paper no. SO-229. New Orleans: Southern Forest Experiment Station.

Black, Richard. 1990. " 'Regional Political Ecology' in Theory and Practice: A Case Study from Northern Portugal." *Transactions of the Institute of British Geographers* 15:35–47.

Black, Robert, Nuchnart Jonglaekha, and Vijit Thanormthin. 1987. "Problems Concerning Pesticide Use in Highland Agriculture, Northern Thailand." In Joyce Tait and Banpot Napompeth, eds., *Management of Pests and Pesticides*, 28–37. Boulder, Colo.: Westview.

Blaikie, Piers. 1985. *The Political Economy of Soil Erosion in Developing Countries*. London: Longman.

——. 1995a. "Understanding Environmental Issues." In Stephen Morse and Michael Stocking, eds., *People and Environment*, 1–30. London: UCL Press.

——. 1995b. "Changing Environments or Changing Views? A Political Ecology for Developing Countries." *Geography* 80:203–14.

Blaikie, Piers, and Harold Brookfield. 1987a. "Defining and Debating the Problem." In Piers Blaikie and Harold Brookfield, eds., *Land Degradation and Society*, 1–26. London: Methuen.

——. 1987b. "Approaches to the Study of Land Degradation." In Piers Blaikie and Harold Brookfield, eds., *Land Degradation and Society*, 27–48. London: Methuen.

Boesen, Jannik, and A. T. Mohele. 1979. *The "Success Story" of Peasant Tobacco Production in Tanzania*. Uppsala, Sweden: Scandinavian Institute of African Studies.

Bonanno, Alessandro. 1991. "The Globalization and Theories of the State of the Agricultural and Food Sector." *International Journal of Sociology of Agriculture and Food* 1:15–30.

——. 1992. "The Restructuring of the Agricultural and Food System." *Agriculture and Human Values* 8:72–82.

——. 1994. "The Locus of Polity Action in a Global Setting." In Alessandro Bonanno, Lawrence Busch, William H. Friedland, Lourdes Gouveia, and Enzo Mingione, eds., *From Columbus to ConAgra: The Globalization of Agriculture and Food*, 251–64. Lawrence: University Press of Kansas.

Bonanno, Alessandro, and Douglas Constance. 1996. *Caught in the Net: The Global Tuna Industry, Environmentalism, and the State*. Lawrence: University Press of Kansas.

Bonanno, Alessandro, Lawrence Busch, William H. Friedland, Lourdes Gouveia, and Enzo Mingione. 1994. Introduction to Alessandro Bonanno, Lawrence Busch, William H. Friedland, Lourdes Gouveia, and Enzo Mingione, eds., *From Columbus to ConAgra: The Globalization of Agriculture and Food*, 1–26. Lawrence: University Press of Kansas.

Borrell, Brent, and Sandy Cuthbertson. 1991. *EC Banana Policy, 1992: Picking the Best Option*. Canberra, Australia: Center for International Economics.

Bottrell, D. G. 1984. "Government Influence on Pesticide Use in Developing Countries." *Insect Science Applications* 5:151–55.

Bousquet, E. 1989. "Islands Face Pesticide Alert." *EC News*, March 31–April 1, p. 1.

Braverman, Harry. 1974. *Labor and Monopoly Capital: The Degradation of Work in the Twentieth Century*. New York: Monthly Review.

Brecher, Jeremy, and Tim Costello. 1994. *Global Village or Global Pillage: Economic Restructuring from the Bottom Up*. Boston: South End.

Brierley, John S. 1985. "Idle Land in Grenada: A Review of Its Causes and the PRG's Approach to Reducing the Problem." *Canadian Geographer* 29:298–309.

——. 1987. "Land Fragmentation and Land-Use Patterns in Grenada." In Jean Besson and Janet Momsen, eds., *Land and Development in the Caribbean*, 194–209. London: Macmillan.

——. 1988. "A Retrospective on West Indian Small Farming, with an Update from Grenada." In John S. Brierley and Hymie Rubenstein, eds., *Small Farming and Peasant Resources in the Caribbean*, 63–82. Winnipeg, Manitoba: University of Manitoba.

Brierley, John S., and Hymie Rubenstein, eds. 1988. *Small Farming and Peasant Resources in the Caribbean*. Winnipeg: University of Manitoba.

Browne, J. Orde. 1939. *Labour Conditions in the West Indies*, Cmd 6070. London: His Majesty's Stationery Office.

Browne, G. 1985. "Current Status of Root and Tuber Crops Production in St. Vincent." Paper presented at the workshop of the Caribbean Network of Root and Tuber Crops, Guadeloupe, 9–10 July.

Bryant, Raymond L. 1992. "Political Ecology: An Emerging Research Agenda in Third-World Studies." *Political Geography* 11:12–36.

——. 1996. "Romancing Colonial Forestry: The Discourse of 'Forestry as Progress' in British Burma." *Geographical Journal* 162:169–78.

Buch-Hansen, Mogens, and Henrik Marcussen. 1982. "Contract Farming and the Peasantry: Cases from Western Kenya." *Review of African Political Economy* 23:9–36.

Bull, D. 1982. *A Growing Problem: Pesticides and the Third World Poor*. Oxford: Oxfam.

Burawoy, Michael. 1979. *Manufacturing Consent: Changes in the Labor Process under Monopoly Capitalism*. Chicago: University of Chicago Press.

Burbach, Roger, and Patricia Flynn. 1980. *Agribusiness in the Americas*. New York: Monthly Review Press.

Butzer, Karl W. 1989. "Cultural Ecology." In Gary Gaile and Cort J. Willmott, eds., *Geography in America*, 192–208. Columbus, Ohio: Merrill Publishing.

Cameron, Norman. 1967. "Ashton Warner's Account of Slavery." *Flambeau* 8:10–13.

Cargill Technical Services. 1995. *Proposals for Restructuring the Windward Islands Banana Industry*. Surrey: Cargill Technical Services.

Caribbean Agricultural Research and Development Institute (CARDI). 1985. *Risk Assessment of Agrochemicals in the Eastern Caribbean*. Proceedings of the CARDI/UNESCO Workshop, St. Lucia.

Caribbean Food and Nutrition Institute. 1974. *Food Composition Tables for Use in the English-Speaking Caribbean*. Caribbean Food and Nutrition Institute. Kingston, Jamaica.

Caribbean Insight. 1997. "Banana Trade under Fire from WTO Panel." April, 1, 12.

Carmichael, Mrs. 1833. *Domestic Manners and Social Condition of the White, Coloured, and Negro Population of the West Indies*. Vol. 1. London: Whittaker, Treacher.

Carney, Judith. 1988. "Struggles over Crop Rights and Labour within Contract Farming Households in a Gambian Irrigated Rice Project." *Journal of Peasant Studies* 15:334–49.

——. 1992. "Peasant Women and Economic Transformation in The Gambia." *Development and Change* 23:67–90.

——. 1994. "Contracting a Food Staple in The Gambia." In Peter D. Little and Michael J. Watts, eds., *Living under Contract: Contract Farming and Agrarian Transformation in Sub-Saharan Africa*, 167–87. Madison: University of Wisconsin Press.

——. 1996. "Converting the Wetlands, Engendering the Environment: The Intersection of Gender with Agrarian Change in Gambia." In Richard Peet and Michael Watts, eds., *Liberation Ecologies: Environment, Development, Social Movements*, 165–87. London: Routledge.

Carriere, Jean. 1991. "The Crisis in Costa Rica: An Ecological Perspective." In David Goodman and Michael Redclift, eds., *Environment and Development in Latin America*, 184–204. Manchester: Manchester University Press.

Chronicle of the West India Committee. 1960. "U.K. Banana Imports: 1959 Arrivals Beat Pre-War Record," May, 130.

——. 1963. "Cameroons Banana Preference to End," August, 432.

Clapp, Roger. 1988. "Representing Reciprocity, Reproducing Domination: Ideology and the Labour Process in Latin American Contract Farming." *Journal of Peasant Studies* 16:5–39.

——. 1994. "The Moral Economy of the Contract." In Peter D. Little and Michael J. Watts, eds., *Living under Contract: Contract Farming and Agrarian Transformation in Sub-Saharan Africa,* 78–94. Madison: University of Wisconsin Press.

Cloke, Paul, and Richard Le Heron. 1994. "Agricultural Deregulation: The Case of New Zealand." In Philip Lowe, Terry Marsden, and Sarah Whatmore, eds., *Regulating Agriculture,* 104–26. London: David Fulton Publishers.

Collins, Jane L. 1987. "Labor Scarcity and Ecological Change." In Peter Little and Michael Horowitz, eds., *Lands at Risk in the Third World: Local-Level Perspectives,* 19–37. Boulder, Colo.: Westview.

——. 1993. "Gender, Contracts, and Wage Work: Agricultural Restructuring in Brazil's São Francisco Valley." *Development and Change* 24:53–82.

Collymore, Jeremy. 1984. "Agricultural Decisions of Small Farmers in St. Vincent." Master's thesis, University of West Indies, Mona, Jamaica.

Colthurst, John. 1977. *The Colthurst Journal: Journal of a Special Magistrate in the Islands of Barbados and St. Vincent, July 1835–September 1838.* Edited by Woodville K. Marshall. Millwood, N.Y.: KTO Press.

Commission of Enquiry. 1959. "Report of the Commission of Enquiry into the Affairs of the SVBGA." Kingstown, St. Vincent: Government Printing Office.

Commonwealth Economic Committee. 1954. *Fruit: A Review.* London: Her Majesty's Stationery Office.

Compton, John. 1965. "Address Delivered to WINBAN Meeting." *WINBAN News* 1:18–21.

——. 1974. "St. Lucia's Premier Opens WINBAN Annual General Meeting." *WINBAN News* 9:9, 37.

Constance, Douglas H., and William D. Heffernan. 1991. "The Global Poultry Agro/Food Complex." *International Journal of Sociology of Agriculture and Food* 1:126–42.

Constance, Douglas H., Alessandro Bonanno, and William D. Heffernan. 1995. "Global Contested Terrain: The Case of the Tuna-Dolphin Controversy." *Agriculture and Human Values* 12:19–33.

Cook, Ian. 1994. "New Fruits and Vanity: Symbolic Production in the Global Food Economy." In Alessandro Bonanno, Lawrence Busch, William H. Friedland, Lourdes Gouveia, and Enzo Mingione, eds., *From Columbus to ConAgra: The Globalization of Agriculture and Food,* 232–48. Lawrence: University Press of Kansas.

Cowen, Michael. 1986. "Change in State Power, International Conditions, and Peasant Producers: The Case of Kenya." *Journal of Development Studies* 22:355–84.

Cox, Robert W. 1996. "A Perspective on Globalization." In James H. Mittelman, ed., *Globalization: Critical Reflections,* 21–30. Boulder, Colo.: Lynne Reinner.

Crompton, Rosemary, and Stuart Reid. 1982. "The Deskilling of Clerical Work." In Stephen Wood, ed., *The Degradation of Work? Skill, Deskilling, and the Labour Process,* 163–84. London: Hutchinson.

Currie, Kate, and Larry Ray. 1986. "On the Class Location of Contract Farmers in the Kenyan Economy." *Economy and Society* 15:445–75.

Curtin, Philip. 1954. "The British Sugar Duties and West Indian Prosperity." *Journal of Economic History* 14:157–64.

Daddieh, Cyril. 1994. "Contract Farming and Palm Oil Production in Côte d'Ivoire and Ghana." In Peter D. Little and Michael J. Watts, eds., *Living under Contract: Contract Farming and Agrarian Transformation in Sub-Saharan Africa*, 188–215. Madison: University of Wisconsin Press.

Davies, Peter. 1990. *Fyffes and the Banana Musa sapientum: A Centenary History, 1888–1988*. London: Athlone Press.

Davis, John Emmeus. 1980. "Capitalist Agricultural Development and the Exploitation of the Propertied Laborer." In Frederick H. Buttel and Howard Newby, eds., *The Rural Sociology of Advanced Societies: Critical Perspectives*, 133–53. Montclair, N.J.: Allanheld, Osmun.

Davy, John. 1854. *The West Indies, before and since Slave Emancipation, Comprising the Windward and Leeward Islands' Military Command*. London: Cash.

Demacque, David. 1983. "Bananas and the Eastern Caribbean: The Case of St. Lucia." Campaign against World Poverty, London.

Dinham, Barbara, and Colin Hines. 1984. *Agribusiness in Africa*. Trenton, N.J.: Africa World Press.

Duncan, Ebenezer. 1970. *A Brief History of St. Vincent with Studies in Citizenship*. 5th ed. Kingstown, St. Vincent: Vincentian.

Durant-Gonzalez, Victoria. 1985. "Higglering: Rural Women and the Internal Market System in Jamaica." In P. I. Gomes, ed., *Rural Development in the Caribbean*, 103–22. London: C. Hurst.

Durham, William H. 1995. "Political Ecology and Environmental Destruction in Latin America." In Michael Painter and William H. Durham, eds., *The Social Causes of Environmental Destruction in Latin America*, 249–64. Ann Arbor: University of Michigan Press.

Ecumenical Committee for Corporate Responsibility. 1994. *Going Bananas*. Fareham, United Kingdom: Ecumenical Committee for Corporate Responsibility.

Ellis, Frank. 1975. "An Institutional Approach to Tropical Commodity Trade: Case-Study of Banana Exports from the Commonwealth Caribbean." Report for Commonwealth Secretariat, London.

Engledow, F. L. 1945. *West India Royal Commission: Report on Agriculture, Fisheries, Forestry, and Veterinary Matters*. Cmd. 6608. London: His Majesty's Stationery Office.

European Commission. 1995. "Report on the Operation of the Banana Regime." Brussels: European Commission.

Fagoonee, I. 1987. "Pesticide Practice among Vegetable Growers in Mauritius." In Joyce Tait and Banpot Napompeth, eds., *Management of Pests and Pesticides*, 175–81. Boulder, Colo.: Westview.

Felton, Mark. 1981a. "The Long Term Supply Function for Bananas in the Windward Islands." Report for WINBAN, Castries, St. Lucia.

——. 1981b. "A Report on Equipment for Banana Farmers in the Windward Islands." Report for WINBAN, Castries, St. Lucia.

Fentem, Arlin. 1961. *Commercial Geography of St. Vincent*. Bloomington: Department of Geography, Indiana University.

Finkel, Herman J. 1964. "Patterns of Land Tenure in the Leeward and Windward Islands and Their Relevance to Problems of Agricultural Development in the West Indies." *Economic Geography* 40:163–72.

Food and Agriculture Organization (FAO). 1995. *FAO Yearbook: Trade*. Rome: FAO.

——. 1996. *FAO Yearbook: Trade*. Rome: FAO.

Franke, Richard W., and Barbara H. Chasin. 1980. *Seeds of Famine: Ecological Destruction and the Development Dilemma in the West African Sahel*. Montclair, N.J.: Allenheld, Osmun.

Fraser, Adrian. 1980. "Development of a Peasantry in St. Vincent: 1846–1912." Master's thesis, University of the West Indies.

——. 1986. "Peasants and Agricultural Labourers in St. Vincent and the Grenadines: 1899–1951." Ph.D. diss., University of Western Ontario.

Friedmann, Harriet. 1991. "Changes in the International Division of Labor: Agri-Food Complexes and Export Agriculture." In William H. Friedland, Lawrence Busch, Frederick H. Buttel, and Allan P. Rudy, eds., *Towards a New Political Economy of Agriculture*, 65–93. Boulder, Colo.: Westview.

Friedmann, Harriet, and Philip McMichael. 1989. "Agriculture and the State System." *Sociologia Ruralis* 29:93–117.

Fröbel, Folker, Jurgen Heinrichs, and Otto Kreye. 1980. *The New International Division of Labor*. Cambridge: Cambridge University Press.

Frucht, Richard. 1967. "A Caribbean Social Type: Neither 'Peasant' nor 'Proletarian.'" *Social and Economic Studies* 16:295–301.

George, Susan. 1977. *How the Other Half Dies: The Real Reasons for World Hunger*. Montclair, N.J.: Allanheld, Osmun.

Gereffi, Gary. 1996. "The Elusive Last Lap in the Quest for Developed-Country Status." In James H. Mittelman, ed., *Globalization: Critical Reflections*, 53–81. Boulder, Colo.: Lynne Reinner.

Gertler, Meric S. 1992. "Flexibility Revisited: Districts, Nation-States, and the Forces of Production." *Transactions of the Institute of British Geographers* 17:259–78.

Glover, David. 1984. "Contract Farming and Smallholder Outgrower Schemes in Less-Developed Countries." *World Development* 12:1143–57.

Glover, David, and Ken Kusterer. 1990. *Small Farmers, Big Business: Contract Farming and Rural Development*. New York: St. Martin's.

Goe, W. Richard, and Martin Kenney. 1991. "The Restructuring of the Global Economy and the Future of U.S. Agriculture." In Kenneth Pigg, ed., *The Future of Rural America*, 137–55. Boulder, Colo.: Westview.

Goldberg, Karen A. 1985. "Efforts to Prevent Misuse of Pesticides Exported to Developing Countries." *Ecology Law Quarterly* 12:1025–51.

Goldman, Abraham. 1986. "Pest Hazards and Pest Management by Small Scale Farmers in Kenya." Ph.D. diss., Clark University.

Goldsmith, Arthur. 1985. "The Private Sector and Rural Development: Can Agribusiness Help the Small Farmer?" *World Development* 13:1125–38.

Gonzalez, Nancy. 1988. *Sojourners of the Caribbean: Ethnogenesis and Ethnohistory of the Garifuna*. Urbana: University of Illinois Press.

Gooding, E. G. B., ed. 1980. *Pests and Pesticide Management in the Caribbean*. Bridgetown, Barbados: Consortium for International Crop Protection.

Goodman, David. 1990. "Farming and Biotechnology: New Approaches to Rural Development." In Henry Buller and Susan Wright, eds., *Rural Development: Problems and Practices*, 97–108. Aldershot: Avebury.

Goodman, David, and Michael Redclift. 1991. *Refashioning Nature: Food, Ecology, and Culture*. London: Routledge.

Goodman, David, and Michael Watts. 1994. "Reconfiguring the Rural or Fording the Divide?: Capitalist Restructuring and the Global Agro-Food System." *Journal of Peasant Studies* 22:1–49.

Grint, Keith. 1991. *The Sociology of Work*. Oxford: Polity.

Grossman, Lawrence S. 1977. "Man-Environment Relations in Anthropology and Geography." *Annals of the Association of American Geographers* 67:126–44.

——. 1984a. *Peasants, Subsistence Ecology, and Development in the Highlands of Papua New Guinea*. Princeton, N.J.: Princeton University Press.

——. 1984b. "Collecting Time-use Data in Third World Rural Communities." *Professional Geographer* 36:444–54.

——. 1992a. "Pesticides, People, and the Environment in St. Vincent." *Caribbean Geography* 3:175–86.

——. 1992b. "Pesticides, Caution, and Experimentation in St. Vincent, Eastern Caribbean." *Human Ecology* 20:53–74.

——. 1993. "The Political Ecology of Banana Exports and Local Food Production in St. Vincent, Eastern Caribbean." *Annals of the Association of American Geographers* 83:347–67.

——. 1994. "British Aid and Windwards Bananas: The Case of St. Vincent and the Grenadines." *Social and Economic Studies* 43:151–79.

——. Forthcoming. "Diet, Income, and Agriculture in an Eastern Caribbean Village." *Human Ecology*.

Guan-Soon, Lim, and Ong Seng-Hock. 1987. "Environmental Problems of Pesticide Usage in Malaysian Rice Fields—Perceptions and Future Considerations." In Joyce Tait and Banpot Napompeth, eds., *Management of Pests and Pesticides*, 10–21. Boulder, Colo.: Westview.

Guillet, David. 1981. "Surplus Extraction, Risk Management, and Economic Change among Peruvian Peasants." *Journal of Development Studies* 18:3–24.

Hall, Douglas. 1978. "The Flight from the Estates Reconsidered: The British West Indies, 1838–1842." *Journal of Caribbean History* 10/11:7–24.

Handler, J. S. 1971. "The History of Arrowroot and the Origin of Peasantries in the British West Indies." *Journal of Caribbean History* 2:46–93.

Hanseen, Linda Irene. 1990. "The Dynamics of Peasant Marketing in Dominica, West Indies." Ph.D. diss., State University of New York at Stony Brook.

Hardesty, Donald. 1977. *Ecological Anthropology*. New York: John Wiley.

Hart, Ansell. 1954. "The Banana in Jamaica: Export Trade." *Social and Economic Studies* 3:212–29.

Hecht, Susanna B. 1985. "Environment, Development, and Politics: Capital Accumulation and the Livestock Sector in Eastern Amazonia." *World Development* 13:663–84.

Henderson, T. H., Everold Hosein, Dunstan Campbell, Roop Dass, David Dolly, and

Justin Francis. 1975. "Constraints to the Adoption of Improved Practices in the Windward Islands Banana Industry." University of the West Indies, St. Augustine, Trinidad.

Hershkovitz, Linda. 1993. "Political Ecology and Environmental Management in the Loess Plateau, China." *Human Ecology* 21:327–53.

Higman, B. W. 1984. *Slave Populations of the British Caribbean, 1807–1834*. Baltimore: Johns Hopkins University Press.

Hirst, Paul, and Jonathan Zeitlin. 1992. "Flexible Specialization versus Post-Fordism." In Michael Storper and Allen J. Scott, eds., *Pathways to Industrialization and Regional Development*, 70–115. London: Routledge.

Howard, Richard A., and Elizabeth S. Howard. 1983. *Alexander Anderson's Geography and History of St. Vincent, West Indies*. London: Linnean Society of London.

Hyden, Goran. 1980. *Beyond Ujamaa in Tanzania: Underdevelopment and an Uncaptured Peasantry*. Berkeley: University of California Press.

Imperial Economic Committee. 1926. *Report of the Imperial Economic Committee on Marketing and Preparing for Market of Foodstuffs Produced in the Overseas Parts of the Empire: Third Report—Fruit*. Cmd. 2658. London: His Majesty's Stationery Office.

Isaacs, Philmore. 1989. "Pesticides Used in St. Vincent and the Grenadines." Paper presented at the FAO workshop "Safe and Effective Use of Pesticides," Trinidad.

Jackson, Jeremy C., and Angela P. Cheater. 1994. "Contract Farming in Zimbabwe: Case Studies of Sugar, Tea, and Cotton." In Peter D. Little and Michael J. Watts, eds., *Living under Contract: Contract Farming and Agrarian Transformation in Sub-Saharan Africa*, 140–66. Madison: University of Wisconsin Press.

Jaffee, Steven. 1994. "Contract Farming in the Shadow of Competitive Markets: The Experience of Kenyan Horticulture." In Peter D. Little and Michael J. Watts, eds., *Living under Contract: Contract Farming and Agrarian Transformation in Sub-Saharan Africa*, 97–139. Madison: University of Wisconsin Press.

James, Canute. 1997. "Clinton in Pledge on Caribbean Exports." *Financial Times* (London), 13 May, 4.

Jarosz, Lucy. 1996a. "Working in the Global Food System: A Focus for International Comparative Analysis." *Progress in Human Geography* 20:41–55.

——. 1996b. "Defining Deforestation in Madagascar." In Richard Peet and Michael Watts, eds., *Liberation Ecologies: Environment, Development, Social Movements*, 148–64. London: Routledge.

Jessop, Bob. 1992. "Fordism and Post-Fordism: A Critical Reformulation." In Michael Storper and Allen J. Scott, eds., *Pathways to Industrialization and Regional Development*, 46–69. London: Routledge.

Jim Fitzpatrick and Associates. 1990. *Trade Policy and the EC Banana Market: An Economic Analysis*. Dublin: Jim Fitzpatrick and Associates.

John, Karl E. V. 1974. "Policies and Programmes of Intervention into the Agrarian Structure of St. Vincent: 1890–1974." Master's thesis, University of Waterloo.

Johnson, Alan. 1972. "Individuality and Experimentation in Traditional Agriculture." *Human Ecology* 1:149–59.

Jolly, A. L. 1947. "Foreword: Preliminary Examination of the Economic and Fiscal Structure of St. Vincent." In Bernard Gibbs, ed., *A Plan of Development for the*

Colony of St. Vincent, Windward Islands, British West Indies, ix–xviii. Port-of-Spain, Trinidad: Guardian Commercial Printers.

Kaplinsky, Raphael. 1988. "Restructuring the Capitalist Labour Process: Some Lessons from the Car Industry." *Cambridge Journal of Economics* 12:451–70.

Katzin, Margaret. 1960. "The Business of Higglering in Jamaica." *Social and Economic Studies* 9:297–331.

Kennedy, Eileen, and Howarth E. Bouis. 1993. *Linkages between Agriculture and Nutrition: Implications for Policy and Research*. Washington, D.C.: International Food Policy Research Institute.

Kenney, Martin, Linda M. Loboa, James Curry, and W. Richard Goe. 1989. "Midwestern Agriculture in US Fordism." *Sociologia Ruralis* 29:131–48.

——. 1991. "Agriculture in U.S. Fordism: The Integration of the Productive Consumer." In William H. Friedland, Lawrence Busch, Frederick H. Buttel, and Allan P. Rudy, eds., *Towards a New Political Economy of Agriculture*, 173–88. Boulder, Colo.: Westview.

Kim, Chul-Kyoo, and James Curry. 1993. "Fordism, Flexible Specialization, and Agri-Industrial Restructuring." *Sociologia Ruralis* 33:61–80.

Kjekshus, H. 1977. *Ecology Control and Economic Development in East Africa: The Case of Tanganyika*. Berkeley: University of California Press.

Koc, Mustafa. 1994. "Globalization as a Discourse." In Alessandro Bonanno, Lawrence Busch, William H. Friedland, Lourdes Gouveia, and Enzo Mingione, eds., *From Columbus to ConAgra: The Globalization of Agriculture and Food*, 265–80. Lawrence: University Press of Kansas.

Korovkin, Tanya. 1992. "Peasants, Grapes, and Corporations: The Growth of Contract Farming in a Chilean Community." *Journal of Peasant Studies* 19:228–54.

Labrianidis, L. 1995. "Flexibility in Production through Subcontracting: The Case of the Poultry Meat Industry in Greece." *Environment and Planning A* 27:193–209.

Lappé, Frances Moore, and Joseph Collins. 1977. *Food First: Beyond the Myth of Scarcity*. New York: Ballantine Books.

Laramee, Peter. 1975. "Problems of Small Farmers under Contract Marketing, with Special Reference to a Case in Chiengmai Province, Thailand." *Economic Bulletin for Asia and the Pacific* 26:43–57.

Leach, Melissa, and Robin Mearns, eds. 1996. *The Lie of the Land: Challenging Received Wisdom on the African Environment*. Oxford: James Currey.

Lee, David. 1982. "Beyond Deskilling: Skill, Craft, and Class." In Stephen Wood, ed., *The Degradation of Work? Skill, Deskilling, and the Labour Process*, 146–62. London: Hutchison.

LeFranc, E. R. 1993. *Rural Land Tenure Systems in St. Lucia*. Mona, Jamaica: Institute of Social and Economic Research.

Lehmann, David. 1986. "Two Paths of Agrarian Capitalism, or a Critique of Chayanovian Marxism." *Comparative Studies in Society and History* 28:601–27.

Lipietz, Alain. 1986. "New Tendencies in the International Division of Labor: Regimes of Accumulation and Modes of Regulation." In Allen J. Scott and Michael Storper, eds., *Production, Work, Territory*, 16–40. Boston: Allen and Unwin.

——. 1987. *Mirages and Miracles: The Crises of Global Fordism*. London: Verso.

———. 1992. *Towards a New Economic Order: Postfordism, Ecology, and Democracy*. New York: Oxford University Press.

Little, Peter D. 1994. "Contract Farming and the Development Question." In Peter D. Little and Michael J. Watts, eds., *Living under Contract: Contract Farming and Agrarian Transformation in Sub-Saharan Africa*, 216–47. Madison: University of Wisconsin Press.

Little, Peter D., and Michael J. Watts. 1994. Introduction to Peter D. Little and Michael J. Watts, eds., *Living under Contract: Contract Farming and Agrarian Transformation in Sub-Saharan Africa*, 3–18. Madison: University of Wisconsin Press.

Llambi, Luis. 1990. "Transitions to and within Capitalism." *Sociologia Ruralis* 30:174–96.

———. 1994. "Comparative Advantages and Disadvantages in Latin American Nontraditional Fruit and Vegetable Exports." In Philip McMichael, ed., *The Global Restructuring of Agro-Food Systems*, 190–213. Ithaca: Cornell University Press.

Long, Frank. 1982. "The Food Crisis in the Caribbean." *Third World Quarterly* 4:758–70.

Lowe, Philip, Terry Marsden, and Sarah Whatmore. 1994. "Changing Regulatory Orders: The Analysis of the Economic Governance of Agriculture." In Philip Lowe, Terry Marsden, and Sarah Whatmore, eds., *Regulating Agriculture*, 1–30. London: David Fulton Publishers.

Lowenthal, David. 1961. "Caribbean Views of Caribbean Land." *Canadian Geographer* 5:1–9.

McAfee, Kathy. 1991. *Storm Signals: Structural Adjustment and Development Alternatives in the Caribbean*. London: Zed.

McConnie, H. S. N.d. *A Guide to the Cultivation of St. Vincent Superfine Sea Island Cotton*. Camden Park Experiment Station Bulletin no. 1. Kingstown, St. Vincent: Department of Agriculture.

McElroy, J., and K. Albuquerque. 1990. "Sustainable Small-Scale Agriculture in Small Caribbean Islands." *Society and Natural Resources* 3:109–29.

McGee, Terence. 1978. "Western Geography and the Third World." *American Behavioral Scientist* 22:93–114.

McIntosh, C. E., and P. Manchew. 1985. "Nutritional Needs, Food Availability, and the Realism of Self-Sufficiency." In P. Gomes, ed., *Rural Development in the Caribbean*, 212–31. London: Hurst.

McMichael, Philip. 1991. "Food, the State, and the World Economy." *International Journal of Sociology of Agriculture and Food* 1:71–85.

———. 1992. "Tensions between National and International Control of the World Food Order." *Sociological Perspectives* 35:343–65.

———. 1993. "Agro-Food Restructuring in the Pacific Rim." In Ravi Palat, ed., *Pacific-Asia and the Future of the World-System*, 103–16. Westport: Greenwood.

———. 1994a. "Global Restructuring: Some Lines of Inquiry." In Philip McMichael, ed., *The Global Restructuring of Agro-Food Systems*, 277–300. Ithaca, N.Y.: Cornell University Press.

———. 1994b. "Introduction: Agro-Food System Restructuring—Unity in Diversity." In Philip McMichael, ed., *The Global Restructuring of Agro-Food Systems*, 1–17. Ithaca, N.Y.: Cornell University Press.

———. 1994c. "GATT, Global Regulation, and the Construction of a New Hegemonic

Order." In Philip Lowe, Terry Marsden, and Sarah Whatmore, eds., *Regulating Agriculture*, 163–90. London: David Fulton Publishers.

McMichael, Philip, and Chul-Kyoo Kim. 1994. "Japanese and South Korean Agricultural Restructuring in Comparative and Global Perspective." In Philip McMichael, ed., *The Global Restructuring of Agro-Food Systems*, 21–52. Ithaca, N.Y.: Cornell University Press.

McMichael, Philip, and David Myhre. 1991. "Global Regulation vs. the Nation-State: Agro-Food Systems and the New Politics of Capital." *Capital and Class* 43:83–105.

Mann, Susan A. 1990. *Agrarian Capitalism in Theory and Practice*. Chapel Hill: University of North Carolina Press.

Mann, Susan A., and James M. Dickenson. 1978. "Obstacles to the Development of a Capitalist Agriculture." *Journal of Peasant Studies* 5:466–81.

Marie, J. M. 1979. *Agricultural Diversification in a Small Economy—the Case for Dominica*. Cave Hill, Barbados: University of the West Indies.

Marsden, Terry. 1996. "Rural Geography Trend Report: The Social and Political Bases of Rural Restructuring." *Progress in Human Geography* 20:246–58.

Marshall, Woodville K. 1965. "Social and Economic Problems in the Windward Islands, 1838–65." In F. M. Andre and T. G. Mathews, eds., *The Caribbean in Transition*, 234–57. Rio Piedras, Puerto Rico: Institute of Caribbean Studies.

——. 1968. "Notes on Peasant Development in the West Indies since 1838." *Social and Economic Studies* 17:252–63.

——. 1991. "Provision Ground and Plantation Labour in Four Windward Islands: Competition for Resources during Slavery." *Slavery and Abolition* 12:48–67.

Matthew, Cyril. 1983. "An Overview of the Industry." *Weekend Voice: WINBAN 25th Anniversary Supplement*, 21–22. Castries, St. Lucia.

——. 1984. "Strategies and Policies." WINBAN, Castries, St. Lucia.

——. 1992. "Productivity in the Windward Islands Banana Industry." *WINBAN Newsletter* 2:7.

Mbilinyi, Marjorie. 1988. "Agribusiness and Women Peasants in Tanzania." *Development and Change* 19:549–83.

Medina, Charito P. 1987. "Pest Control Practices and Pesticide Perceptions of Vegetable Farmers in Loo Valley, Benguet, Philippines." In Joyce Tait and Banpot Napompeth, eds., *Management of Pests and Pesticides*, 150–57. Boulder, Colo.: Westview.

Ministry of Agriculture, Fisheries, and Food (MAFF). 1990. "Summary of UK Banana Policy." Document 2J16. London, United Kingdom, February.

Minot, Nicholas. 1986. *Contract Farming and Its Effect on Small Farmers in Less Developed Countries*. Michigan State University Development Papers, Working Paper no. 31. East Lansing: Michigan State University.

Mintz, Sidney W. 1961. "The Question of Caribbean Peasantries: A Comment." *Caribbean Studies* 1:31–34.

——. 1973. "A Note on the Definition of Peasantries." *Journal of Peasant Studies* 1:91–106.

——. 1974. "The Rural Proletariat and the Problem of Rural Proletarian Consciousness." *Journal of Peasant Studies* 1:291–325.

——. 1983. "Reflections on Caribbean Peasantries." *Nieuwe West-Indische Gids* 57:1–17.

——. 1985. "From Plantations to Peasantries in the Caribbean." In Sidney W. Mintz and Sally Price, eds., *Caribbean Contours*, 127–54. Baltimore: Johns Hopkins University Press.

——. 1989. *Caribbean Transformations*. New York: Columbia University Press.

Mintz, Sidney W., and Douglas Hall. 1960. *The Origins of the Jamaican Internal Marketing System*. Publications in Anthropology, no. 57. New Haven: Yale University.

Mintz, Sidney W., and Richard Price. 1992. *The Birth of African-American Culture: An Anthropological Perspective*. Boston: Beacon.

Mittelman, James H. 1996a. "The Dynamics of Globalization." In James H. Mittelman, ed., *Globalization: Critical Reflections*, 1–19. Boulder, Colo.: Lynne Reinner.

——. 1996b. "How Does Globalization Really Work?" In James H. Mittelman, ed., *Globalization: Critical Reflections*, 229–41. Boulder, Colo.: Lynne Reinner.

——, ed. 1996. *Globalization: Critical Reflections*. Boulder, Colo.: Lynne Reinner.

Moberg, Mark A. 1991. "Marketing Policy and the Loss of Food Self-Sufficiency in Rural Belize." *Human Organization* 50:16–25.

Mohan, V. C. 1987. "The Pesticide Dilemma in Malaysia." In Joyce Tait and Banpot Napompeth, eds., *Management of Pests and Pesticides*, 71–78. Boulder, Colo.: Westview.

Momsen, Janet. 1987a. "Land Settlement as an Imposed Solution." In Jean Besson and Janet Momsen, eds., *Land and Development in the Caribbean*, 46–69. London: Macmillan.

——. 1987b. "The Feminization of Agriculture in the Caribbean." In Janet Momsen and Janet Townsend, eds., *Geography of Gender in the Third World*, 344–47. Albany: State University of New York Press.

——. 1988. "Changing Gender Roles in Caribbean Peasant Agriculture." In John S. Brierley and Hymie Rubenstein, eds., *Small Farming and Peasant Resources in the Caribbean*, 83–99. Winnipeg: University of Manitoba.

——. 1992. "Canada-Caribbean Relations: Wherein the Special Relationship?" *Political Geography* 11:501–13.

——. 1993. "Development and Gender Divisions of Labour in the Rural Eastern Caribbean." In Janet Momsen, ed., *Women and Change in the Caribbean*, 232–46. Kingston, Jamaica: Ian Randle.

Moore, Donald S. 1996. "Marxism, Culture, and Political Ecology: Environmental Struggles in Zimbabwe's Eastern Highlands." In Richard Peet and Michael Watts, eds., *Liberation Ecologies: Environment, Development, Social Movements*, 125–47. London: Routledge.

Morvaridi, Behrooz. 1995. "Contract Farming and Environmental Risk: The Case of Cyprus." *Journal of Peasant Studies* 23:30–45.

Moulaert, F., and E. A. Swyngedouw. 1989. "A Regulation Approach to the Geography of Flexible Production Systems." *Environment and Planning D* 7:327–45.

Mourillon, V. J. 1978. *The Dominica Banana Industry from Inception to Independence, 1928–1978*. Roseau, Dominica: Tropical Printers.

MSDS Reference for Crop Protection Chemicals, 1989/90. 1989. Paris: Chemical and Pharmaceutical Press.

Murray, Douglas L. 1994. *Cultivating Crisis: The Human Cost of Pesticides in Latin America*. Austin: University of Texas Press.

Murray, Douglas L., and Polly Hoppin. 1992. "Recurring Contradictions in Agrarian Development: Pesticide Problems in Caribbean Basin Nontraditional Agriculture." *World Development* 20:597–608.

Netting, Robert. 1986. *Cultural Ecology*. 2d ed. Prospect Heights, Ill.: Waveland Press.

Nietschmann, B. 1973. *Between Land and Water: The Subsistence Ecology of the Miskito Indians*. New York: Seminar Press.

Nurse, Keith, and Wayne Sandiford. 1995. *Windward Islands Bananas: Challenges and Options under the Single European Market*. Kingston, Jamaica: Friedrich Ebert Stiftung.

Ó hUallacháin, Breandán. 1996. "Vertical Integration in American Manufacturing: Evidence for the 1980s." *Professional Geographer* 48:343–56.

Page, Brian, and Richard Walker. 1994. "Staple Lessons: Agriculture, Resource Industrialization, and Economic Geography." Paper presented at the Harold Innis Symposium, University of Toronto, Toronto, 24 September.

Painter, Michael. 1995. "Upland-Lowland Production Linkages and Land Degradation in Bolivia." In Michael Painter and William H. Durham, eds., *The Social Causes of Environmental Destruction in Latin America*, 133–68. Ann Arbor: University of Michigan Press.

Panitch, Leo. 1996. "Rethinking the Role of the State." In James H. Mittelman, ed., *Globalization: Critical Reflections*, 83–113. Boulder, Colo.: Lynne Reinner.

Patterson, Karen. 1996. "The Political Ecology of Nontraditional Agricultural Exports and an IPM Project in Jamaica." Master's thesis, Virginia Polytechnic Institute and State University.

Peet, Richard, and Michael Watts. 1996. "Liberation Ecology: Development, Sustainability, and Environment in an Age of Market Triumphalism." In Richard Peet and Michael Watts, eds., *Liberation Ecologies: Environment, Development, Social Movements*, 1–45. London: Routledge.

———, eds. 1996. *Liberation Ecologies: Environment, Development, Social Movements*. London: Routledge.

Persaud, Bishnodat. 1981. "The Export Marketing Arrangements of the Windward Islands Banana Industry." Commonwealth Secretariat, London.

Phillips, C. A., and I. T. Twyford. 1965. "A Quick Look at Martinique's Bananas." *WINBAN News* 1:14–17.

Piore, Michael J., and Charles F. Sabel. 1984. *The Second Industrial Divide: Possibilities for Prosperity*. New York: Basic.

Polius, F., and B. Laville. 1988. "Some Observations on the Effect of High Ratoon Densities on the Growth and Performance of Bananas." WINBAN, Castries, St. Lucia.

Pollard, H. J. 1981. "Food Crop Production for Trinidad's Home Market: An Unfulfilled Potential." *Singapore Journal of Tropical Geography* 2:91–100.

Popenoe, Wilson. 1941. "Banana Culture around the Caribbean." *Tropical Agriculture* 18:8–12, 33–38.

Porter, Gina, and Kevin Phillips-Howard. 1996. "Tensions and Transformation in a Tea Enterprise: Transkei, South Africa." *Geographical Journal* 162:287–94.

——. 1997. "Comparing Contracts: An Evaluation of Contract Farming Schemes in Africa." *World Development* 25:227–38.

Porter, Philip W. 1978. "Geography as Human Ecology." *American Behavioral Scientist* 22:15–39.

——. 1979. *Food and Development in the Semi-Arid Zone of East Africa.* Syracuse: Maxwell School of Citizenship and Public Affairs, Syracuse University.

Pugliese, Enrico. 1991. "Agriculture and the New Division of Labor." In William H. Friedland, Lawrence Busch, Frederick H. Buttel, and Allan P. Rudy, eds., *Towards a New Political Economy of Agriculture,* 137–50. Boulder, Colo.: Westview.

Rainey, William E. 1985. "Dominica Banana Rehabilitation Project Pesticide Assessment." Island Resources Foundation, Washington, D.C.

Rama, Ruth. 1985. "Some Effects of the Internationalization of Agriculture on the Mexican Agricultural Crisis." In Steven E. Sanderson, ed., *The Americas in the New International Division of Labor,* 69–94. New York: Holmes and Meier.

Rappaport, Roy. 1968. *Pigs for the Ancestors: Ritual in the Ecology of a New Guinea People.* New Haven: Yale University Press.

Raynolds, Laura. 1994a. "The Restructuring of Third World Agro-Exports: Changing Production Relations in the Dominican Republic." In Philip McMichael, ed., *The Global Restructuring of Agro-Food System,* 214–37. Ithaca, N.Y.: Cornell University Press.

——. 1994b. "Institutionalizing Flexibility: A Comparative Analysis of Fordist and Post-Fordist Models of Third World Agro-Export Production." In Gary Gereffi and Miguel Korzeniewicz, eds., *Commodity Chains and Global Capitalism,* 143–61. London: Praeger.

Raynolds, Laura, David Myhre, Philip McMichael, Viviana Carro-Figueroa, and Fredrick H. Buttel. 1993. "The 'New' Internationalization of Agriculture: A Reformulation." *World Development* 21:1101–21.

Read, Robert. 1994. "The EC Internal Banana Market: The Issues and the Dilemma." *World Economy* 17:219–35.

Richardson, Bonham C. 1989. "Catastrophes and Change on St. Vincent." *National Geographic Research* 5:111–25.

——. 1992. *The Caribbean in the Wider World, 1492–1992: A Regional Geography.* Cambridge: Cambridge University Press.

——. 1997. *Economy and Environment in the Caribbean: Barbados and the Windwards in the Late 1800s.* Gainesville: University Press of Florida.

Rojas, Eduarado. 1984. "Agricultural Land in the Eastern Caribbean." *Land Use Policy* (January): 39–54.

Roobeek, Annemieke J. 1987. "The Crisis in Fordism and the Rise of a New Technological Paradigm." *Futures* 44:129–54.

Rubenstein, Hymie. 1975. "The Utilization of Arable Land in an Eastern Caribbean Valley." *Canadian Journal of Sociology* 1:157–67.

——. 1983. "Remittances and Rural Underdevelopment in the English-Speaking Caribbean." *Human Organization* 42:295–306.

——. 1987. *Coping with Poverty: Adaptive Strategies in a Caribbean Village.* Boulder, Colo.: Westview.
——. 1995. "Mirror for the Other: Marijuana, Multivocality, and the Media in an Eastern Caribbean Country." *Anthropologica* 37:173–206.
Samatar, Abdi. 1993. "Structural Adjustment as Development Strategy? Bananas, Boom, and Poverty in Somalia." *Economic Geography* 69:25–43.
Sanderson, Steven E. 1985a. "The 'New' Internationalization of Agriculture in the Americas." In Steven E. Sanderson, ed., *The Americas in the New International Division of Labor*, 46–68. New York: Holmes and Meier.
——. 1985b. "A Critical Approach to the Americas in the New International Division of Labor." In Steven E. Sanderson, ed., *The Americas in the New International Division of Labor*, 3–25. New York: Holmes and Meier.
——. 1986a. "The Emergence of the 'World Steer': Internationalization and Foreign Domination in Latin American Cattle Production." In F. Lamond Tullis and W. Ladd Hollist, eds., *Food, the State, and International Political Economy: Dilemmas of Developing Countries*, 123–48. Lincoln: University of Nebraska Press.
——. 1986b. *The Transformation of Mexican Agriculture: International Structure and the Politics of Rural Change.* Princeton: Princeton University Press.
Sauer, Matthias. 1990. "Fordist Modernization of German Agriculture and the Future of Family Farms." *Sociologia Ruralis* 30:260–79.
Sayer, Andrew. 1989. "Postfordism in Question." *International Journal of Urban and Regional Research* 13:666–95.
Schmink, Marianne, and Charles H. Wood. 1987. "The 'Political Ecology' of Amazonia." In Peter Little and Michael H. Horowitz, eds., *Lands at Risk in the Third World: Local-Level Perspectives*, 38–57. Boulder, Colo.: Westview.
Schoenberger, E. 1988. "From Fordism to Flexible Accumulation: Technology, Competitive Strategies, and International Location." *Environment and Planning D* 6:245–62.
Schroeder, Richard A., and Krisnawati Suryanata. 1996. "Gender and Class Power in Agroforestry Systems: Case Studies from Indonesia and West Africa." In Richard Peet and Michael Watts, eds., *Liberation Ecologies: Environment, Development, Social Movements*, 188–204. London: Routledge.
Sealy, Theodore, and Herbert Hart. 1984. *Jamaica's Banana Industry*. Kingston, Jamaica: Jamaica Banana Producers Association.
Shephard, Charles. 1971. *An Historical Account of the Island of Saint Vincent.* 1831. Reprint, London: Frank Cass.
Shephard, C. Y. 1945. *Peasant Agriculture in the Leeward and Windward Islands.* Trinidad: Imperial College of Tropical Agriculture.
Shipton, Parker. 1985. "Land, Credit, and Crop Transitions in Kenya: The Luo Response to Directed Development in Nyanza Province." Ph.D. diss., St. John's College, Cambridge University.
Slater, Courtenay, ed. 1996. *Business Statistics of the United States.* Lanham, Md.: Bernan Press.
Smith, M. G. 1965. *The Plural Society in the British West Indies.* Berkeley: University of California Press.

Spector, J. 1967. "E.E.C. and the West Indian Banana Industry." *West Indies Chronicle* (October): 499–500.

Spinelli, Joseph. 1973. "Land Use and Population in St. Vincent, 1763–1960." Ph.D. diss., University of Florida.

Stemman, Roy. 1986. *Geest: 1935–85*. London: Alsace Print and Production.

Stephen, James. 1830. *The Slavery of the British West India Colonies Delineated, as It Exists Both in Law and Practice, and Compared with the Slavery of Other Countries, Ancient and Modern*. Vol. 2. London: Saunders and Benning.

Stonich, Susan. 1993. *"I Am Destroying the Land!" The Political Ecology of Poverty and Environmental Destruction in Honduras*. Boulder, Colo.: Westview.

Storper, Michael, and Allen J. Scott. 1986. "Production, Work, Territory: Contemporary Realities and Theoretical Tasks." In Allen J. Scott and Michael Storper, eds., *Production, Work, Territory*, 3–15. Boston: Allen and Unwin.

——, eds. 1992. *Pathways to Industrialization and Regional Development*. London: Routledge.

Storper, Michael, and Richard Walker. 1989. *The Capitalist Imperative*. Oxford: Basil Blackwell.

Streeton, Paul. 1993. "The Special Problems of Small Countries." *World Development* 21:197–202.

Sutton, Paul. 1984. "From Neo-colonialism to Neo-colonialism: Britain and the EEC in the Commonwealth Caribbean." In Anthony Payne and Paul Sutton, eds., *Dependency under Challenge: The Political Economy of the Commonwealth Caribbean*, 204–37. Manchester: Manchester University Press.

——. 1995. "The 'New Europe' and the Caribbean." *European Review of Latin American and Caribbean Studies* 59:37–57.

——. 1997. "The Banana Regime of the European Union, the Caribbean, and Latin America." *Journal of Interamerican Studies and World Affairs* 39:5–36.

SVBGA. 1966. *St. Vincent Banana Growers' Manual*. Kingstown, St. Vincent: SVBGA.

——. 1968. *Report of the Board of Management and Accounts for the Year Ended 30th September, 1968*. Kingstown, St. Vincent: SVBGA.

——. 1972. *Report of the Board of Management for the Year Ended 30th September, 1972*. Kingstown, St. Vincent: SVBGA.

——. 1975. *Report of the Board of Management and Accounts for the Year Ended 30th September, 1974*. Kingstown, St. Vincent: SVBGA.

——. 1983. "Minutes of Meeting between WINBAN and the Association's Management Held in the St. Vincent Banana Growers' Association Board Room on 22nd August, 1983 at 10:30 a.m." Kingstown, St. Vincent.

——. 1993a. *Annual Report, 1992*. Kingstown, St. Vincent: SVBGA.

——. 1993b. "Quality Specifications for Green Banana Clusters from the Windwards." *Bananas: Bulletin of the St. Vincent Banana Growers' Association* 2:2–3.

——. 1994. *Annual Report, 1993*. Kingstown, St. Vincent: SVBGA.

——. 1995. *1994 Report and Statement of Accounts to the 41st Annual General Meeting*. Kingstown, St. Vincent: SVBGA.

——. 1996a. *1995 Report and Statement of Accounts to the 42nd Annual General Meeting*. Kingstown, St. Vincent: SVBGA.

——. 1996b. "A Review of 1995." Kingstown, St. Vincent.

SVG. 1955. Banana Growers Association Ordinance, 1954. Act no. 44 of 1954. Kingstown, St. Vincent.

——. 1956. *Annual Trade Report for the Year 1955*. Kingstown, St. Vincent: Government Printing Office.

——. 1973. Pesticide Control Act, 1973. Act no. 23 of 1973. Kingstown, St. Vincent.

——. 1978a. Banana Industry Act, 1978. Act no. 10 of 1978. Kingstown, St. Vincent.

——. 1978b. *Census of Agriculture for St. Vincent: 1972–1973*. Antigua: UNDP—ECCM Printery.

——. 1984. Banana (Protection and Quality Control) Act, 1984. Act no. 33 of 1984. Kingstown, St. Vincent.

SVG Agricultural Department. 1940. *Annual Report of the Agricultural Department, St. Vincent, 1938*. Kingstown, St. Vincent: Government Printing Office.

——. 1945. *Annual Report of the Agricultural Department, St. Vincent, 1944*. Kingstown, St. Vincent: Government Printing Office.

——. 1953. *Annual Report of the Agricultural Department, St. Vincent, 1951*. Kingstown, St. Vincent: Government Printing Office.

SVG Central Planning Division. 1992. *St. Vincent and the Grenadines Development Plan, 1991–1995: Balanced Growth and Sustainable Development*. Kingstown, St. Vincent: Ministry of Finance and Planning.

SVG Department of Agriculture. 1959. *Report of the Department of Agriculture, Forestry, and Fisheries, St. Vincent for the Year Ended 31st December, 1957*. Kingstown, St. Vincent: Government Printing Office.

——. 1984. *Report of the Department of Agriculture for the Year 1980*. Kingstown, St. Vincent: Government Printing Office.

SVG Ministry of Agriculture, Industry, and Labour. 1989. *1985–1986 Agricultural Census for St. Vincent and the Grenadines*. Kingstown, St. Vincent: Ministry of Agriculture, Industry, and Labour.

SVG Statistical Office. 1993. *St. Vincent and the Grenadines: 1991—Population and Housing Census Report*. Vol. 2. Kingstown, St. Vincent: Ministry of Finance and Planning.

SVG Statistical Unit. 1995. *Digest of Statistics for the Year 1993, no. 43*. Kingstown, St. Vincent: Central Planning Division.

Tait, Joyce, and Banpot Napompeth, eds. 1987. *Management of Pests and Pesticides*. Boulder, Colo.: Westview.

Thaman, Randy. 1985. "The Poisoning of Paradise: Pesticides, People, Environmental Pollution, and Increasing Dependency in the Pacific Islands." *South Pacific Forum* 1:165–200.

Thomas, Clive Y. 1988. *The Poor and the Powerless: Economic Policy and Change in the Caribbean*. London: Latin American Bureau.

Thomson, Robert. 1987. *Green Gold: Bananas and Dependency in the Eastern Caribbean*. London: Latin American Bureau.

Thrupp, Lori Ann. 1988. "Pesticides and Policies: Approaches to Pest-Control Dilemmas in Nicaragua and Costa Rica." *Latin American Perspectives* 15:37–70.

——. 1990a. "Inappropriate Incentives for Pesticide Use: Agricultural Credit Requirements in Developing Countries." *Agriculture and Human Values* 7:62–69.

——. 1990b. "Entrapment and Escape from Fruitless Insecticide Use: Lessons from the Banana Sector of Costa Rica." *International Journal of Environmental Studies* 36:173–89.

——. 1994. *Challenges in Latin America's Recent Agroexport Boom*. Washington, D.C.: World Resources Institute.

Thurow, Lester C. 1996. *The Future of Capitalism: How Today's Economic Forces Shape Tomorrow's World*. New York: William Morrow and Company.

Ticknell, Adam, and Jamie Peck. 1992. "Accumulation, Regulation, and the Geographies of Post-Fordism: Missing Links in Regulationist Research." *Progress in Human Geography* 16:190–218.

Tomich, Dale W. 1990. *Slavery in the Circuit of Sugar: Martinique and the World Economy, 1830–1848*. Baltimore: Johns Hopkins University Press.

Trouillot, Michel-Rolph. 1988. *Peasants and Capital: Dominica in the World Economy*. Baltimore: Johns Hopkins University Press.

Twyford, I. T. 1967. "The Quality of Windward Islands Bananas in the United Kingdom." *WINBAN News* 3:11–13.

Ufkes, Francis. 1993. "Trade Liberalization, Agro-Food Politics, and the Globalization of Agriculture." *Political Geography* 12:215–31.

United Nations Development Program [UNDP] Physical Planning Project. 1976a. *St. Vincent National Development Plan*. Vol. 2, *Policy and Strategy*. Kingstown, St. Vincent.

——. 1976b. *St. Vincent National Development Plan*. Vol. 1, *Survey and Analysis*. Kingstown, St. Vincent.

University of the West Indies Development Mission. 1969. *The Development Problem in St. Vincent*. Kingston, Jamaica: Institute of Social and Economic Research.

von Braun, Joachim, and Eileen Kennedy. 1986. *Commercialization of Subsistence Agriculture: Income and Nutritional Effects in Developing Countries*. Washington, D.C.: International Food Policy Research Institute.

von Bülow, Dorthe, and Anne Sørensen. 1993. "Gender and Contract Farming: Tea Outgrower Schemes in Kenya." *Review of African Political Economy* 56:38–52.

Walker, Frederick. 1937. "Economic Progress of St. Vincent, B.W.I., since 1927." *Economic Geography* 13:217–34.

Watson, Hilbourne A. 1994. "The United States–Canada Free Trade Agreement, Semiconductors, and a Case Study from Barbados." In Hilbourne A. Watson, ed., *The Caribbean in the Global Political Economy*, 127–46. Boulder, Colo.: Lynne Rienner.

Watts, David. 1987. *The West Indies: Patterns of Development, Culture, and Environmental Change since 1492*. Cambridge: Cambridge University Press.

Watts, Michael. 1983. *Silent Violence: Food, Famine, and Peasantry in Northern Nigeria*. Berkeley: University of California Press.

——. 1990. "Peasants under Contract: Agro-Food Complexes in the Third World." In Henry Bernstein, Ben Crow, Maureen Mackintosh, and Charlotte Martin, eds., *The Food Question: Profits versus People*, 149–62. London: Earthscan.

——. 1992. "Living under Contract: Work, Production Politics, and the Manufacture of Discontent in Peasant Society." In Allan Pred and Michael Watts, eds., *Reworking Modernity*, 65–105. New Brunswick: Rutgers University Press.

——. 1994a. "Life under Contract: Contract Farming, Agrarian Restructuring, and Flexible Accumulation." In Peter D. Little and Michael J. Watts, eds., *Living under Contract: Contract Farming and Agrarian Transformation in Sub-Saharan Africa*, 21–77. Madison: University of Wisconsin Press.

——. 1994b. "Epilogue: Contracting, Social Labor, and Agrarian Transitions." In Peter D. Little and Michael J. Watts, eds., *Living under Contract: Contract Farming and Agrarian Transformation in Sub-Saharan Africa*, 248–57. Madison: University of Wisconsin Press.

——. 1996. "Development III: The Global Agrofood System and Late Twentieth-Century Development (or Kautsky *Redux*)." *Progress in Human Geography* 20:230–45.

Watts, Michael, Peter D. Little, Christopher Mock, Martin Billings, and Steven Jaffee. 1988. *Contract Farming in Africa*. Vol. 1, *Comparative Analysis*. Binghamton, N.Y.: Institute for Development Anthropology.

Wedderburn, Samuel. 1995. *Final Report on the Baseline Survey for the Smallholder Crop Improvement and Marketing Project (St. Vincent and the Grenadines)*. Kingstown, St. Vincent: Ministry of Agriculture, Industry, and Labour.

Weir, David, and Mark Schapiro. 1981. *Circle of Poison: Pesticides and People in a Hungry World*. San Francisco: Institute for Food and Development Policy.

Welch, Barbara. 1994. "Banana Dependency: Albatross or Life Raft for the Windwards?" *Social and Economic Studies* 43:123–49.

——. 1996. *Survival by Association: Supply Management Landscapes of the Eastern Caribbean*. Montreal: McGill-Queen's University Press.

West India Committee Circular. 1948. "Jamaica Banana Industry: Ministry of Food Accepts the Lacatan," September, 187.

——. 1950. "Bananas: Dominica," May, 110.

——. 1951. "The Popularity of Bananas," January, 20.

——. 1956a. "St. Lucia: Banana Industry," April, 101.

——. 1956b. "U.K. Import Duties: Increases on Bananas and Lime Oil," April, 89.

West India Royal Commission. 1897. *Report of the West India Royal Commission*, C.-8655. London: Her Majesty's Stationery Office.

West Indies Chronicle. 1972. "Dollar Imports," October, 412.

Whatmore, Sarah. 1994. "Global Agro-Food Complexes and the Refashioning of Rural Europe." In Ash Amin and Nigel Thrift, eds., *Globalization, Institutions, and Regional Development in Europe*, 46–67. Oxford: Oxford University Press.

——. 1995. "From Farming to Agribusiness: The Global Agro-Food System." In R. J. Johnston, Peter J. Taylor, and Michael J. Watts, eds., *Geographies of Global Change: Remapping the World in the Late Twentieth Century*, 36–49. Oxford: Blackwell.

WIBDECO. 1997. "Economic and Statistical Review, 1996." Castries, St. Lucia: WIBDECO.

Wiley, James. Forthcoming. "Economic Diversification in a Caribbean Mini-State: Dominica in a Neoliberal Era." In Thomas Klak, ed., *Globalization and Neoliberalism: The Caribbean Context*. Boulder, Colo.: Rowman and Littlefield.

Wilson, John. 1986. "The Political Economy of Contract Farming." *Review of Radical Political Economics* 18:47–70.

WINBAN. 1964. *Annual Report of the Windward Islands Banana Growers' Association Limited for Year July, 1963 to June, 1964*. Castries, St. Lucia: WINBAN.

——. 1966. *Annual Report of the Windward Islands Banana Growers' Association for Year July 1965–June 1966*. Castries, St. Lucia: WINBAN.

——. 1967. "U.K. Bid for E.C.M. Membership." *WINBAN News* 3:2–3.

——. 1968. *Annual Report of the Windward Islands Banana Growers' Association for Year July, 1967–June, 1968*. Castries, St. Lucia: WINBAN.

——. 1970. "A Question of Survival." *WINBAN News* 3:2.

——. 1971. "The Economic Importance of the Banana Industry in St. Lucia." *WINBAN News* 4:1, 3, 10.

——. 1973. "Minutes of an Ordinary General Meeting of this Association (WINBAN) Held in St. Lucia on 26th/29th November, 1973." Castries, St. Lucia.

——. 1980a. *Annual Report of the Windward Islands Banana Growers' Association for the Year 1st January to 31st December, 1979*. Castries, St. Lucia: WINBAN.

——. 1980b. "Report on Geest Banana Operations in the West Indies and in the United Kingdom." WINBAN, Castries, St. Lucia.

——. 1981a. *Annual Report of the Windward Islands Banana Growers' Association for the Year 1st January to 31st December, 1980*. Castries, St. Lucia: WINBAN.

——. 1981b. *Banana Growers Manual*. Castries, St. Lucia: WINBAN.

——. 1986a. *WINBAN Annual Report*. Castries, St. Lucia: WINBAN.

——. 1986b. *Banana Growers Manual*. Castries, St. Lucia: WINBAN.

——. 1987. "The Banana in the EEC Market." WINBAN, Castries, St. Lucia.

——. 1990. *Annual Report and Accounts, 1989*. Castries, St. Lucia: WINBAN.

——. 1991. *Windward Islands Banana Growers' Association Annual Report and Accounts, 1990*. Castries, St. Lucia: WINBAN.

——. 1993a. *1992 Annual Report and Accounts*. Castries, St. Lucia: WINBAN.

——. 1993b. *Banana Growers' Manual*. Castries, St. Lucia: WINBAN.

——. 1995. *Quarterly Review of the Windward Islands Banana Industry, January to March, 1995*. Castries, St. Lucia: WINBAN.

Windward Islands Banana Conference. 1965. "Discussions with the Windward Islands Banana Delegation." Report C.O./1080/65.

Wisner, Ben. 1977. "Man-Made Famine in Eastern Kenya." In Phil O'Keefe and Ben Wisner, eds., *Landuse and Development*, 194–215. London: International African Institute.

Wolf, Eric. 1972. "Ownership and Political Ecology." *Anthropological Quarterly* 45:201–5.

"Women Traders of the Caribbean." 1990. In Paul B. Goodwin, ed., *Latin America*, 4th ed., 198–201. Guilford, Conn.: Dushkin Publishing Group.

Wood, E. F. L. 1922. *West Indies: Report by the Honourable E. F. L. Wood, M.P., on His Visit to the West Indies and British Guyana: December, 1921–February, 1922*, Cmd. 1679. London: His Majesty's Stationery Office.

Wood, Stephen. 1982. Introduction to Stephen Wood, ed., *The Degradation of Work? Skill, Deskilling, and the Labour Process*, 11–22. London: Hutchinson.

——. 1987. "The Deskilling Debate, New Technology, and Work Organization." *Acta Sociologica* 30:3–24.
Wood, Stephen, and John Kelly. 1982. "Taylorism, Responsible Autonomy, and Management Strategy." In Stephen Wood, ed., *The Degradation of Work? Skill, Deskilling, and the Labour Process,* 74–89. London: Hutchinson.
Wright, Angus. 1986. "Rethinking the Circle of Poison: The Politics of Pesticide Poisoning among Mexican Farm Workers." *Latin American Perspectives* 13:26–59.
——. 1990. *The Death of Ramon Gonzalez: The Modern Agricultural Dilemma.* Austin: University of Texas Press.
Wright, G. 1929. "Economic Conditions in St. Vincent, B.W.I." *Economic Geography* 5:236–59.
Yapa, Lakshman. 1979. "Ecopolitical Economy of the Green Revolution." *Professional Geographer* 31:371–76.
Zaidi, Iqtidar. 1984. "Farmers' Perception and Management of Pest Hazard." *Insect Science Applications* 5:187–201.
Zimmerer, Karl S. 1991. "Wetland Production and Smallholder Persistence: Agricultural Change in a Highland Peruvian Region." *Annals of the Association of American Geographers* 81:443–63.
——. 1993. "Soil Erosion and Labor Shortages in the Andes with Special Reference to Bolivia, 1953–91: Implications for 'Conservation-with-Development.'" *World Development* 21:1659–75.
——. 1994. "Human Geography and the 'New Ecology': The Prospect and Promise of Integration." *Annals of the Association of American Geographers* 84:108–25.
——. 1996a. "Discourses on Soil Loss in Bolivia: Sustainability and the Search for Socioenvironmental 'Middle Ground.'" In Richard Peet and Michael Watts, eds., *Liberation Ecologies: Environment, Development, Social Movements,* 110–24. London: Routledge.
——. 1996b. "Ecology as Cornerstone and Chimera in Human Geography." In Carville Earle, Kent Mathewson, and Martin S. Kenzer, eds., *Concepts in Human Geography,* 161–88. Lanham, Md.: Rowman and Littlefield.

Index